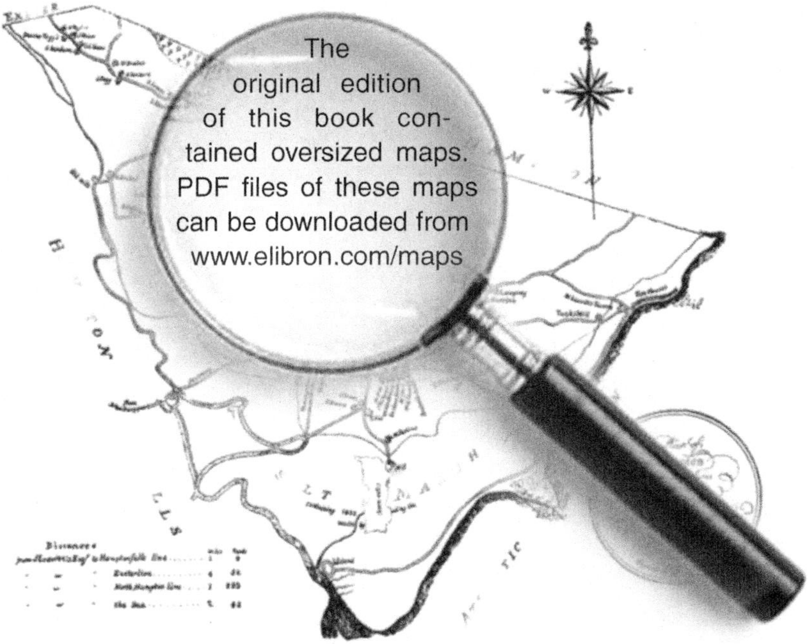
The
original edition
of this book con-
tained oversized maps.
PDF files of these maps
can be downloaded from
www.elibron.com/maps

PROCEEDINGS

OF THE

ASSOCIATION

FOR

PROMOTING THE DISCOVERY

OF THE

INTERIOR PARTS OF AFRICA

Elibron Classics
www.elibron.com

Elibron Classics series.

ISBN 1-4212-3065-8 (paperback)
ISBN 1-4212-3064-X (hardcover)

This Elibron Classics Replica Edition is an unabridged facsimile
of the edition published in 1791,
London.

PROCEEDINGS

OF THE

AFRICAN ASSOCIATION.

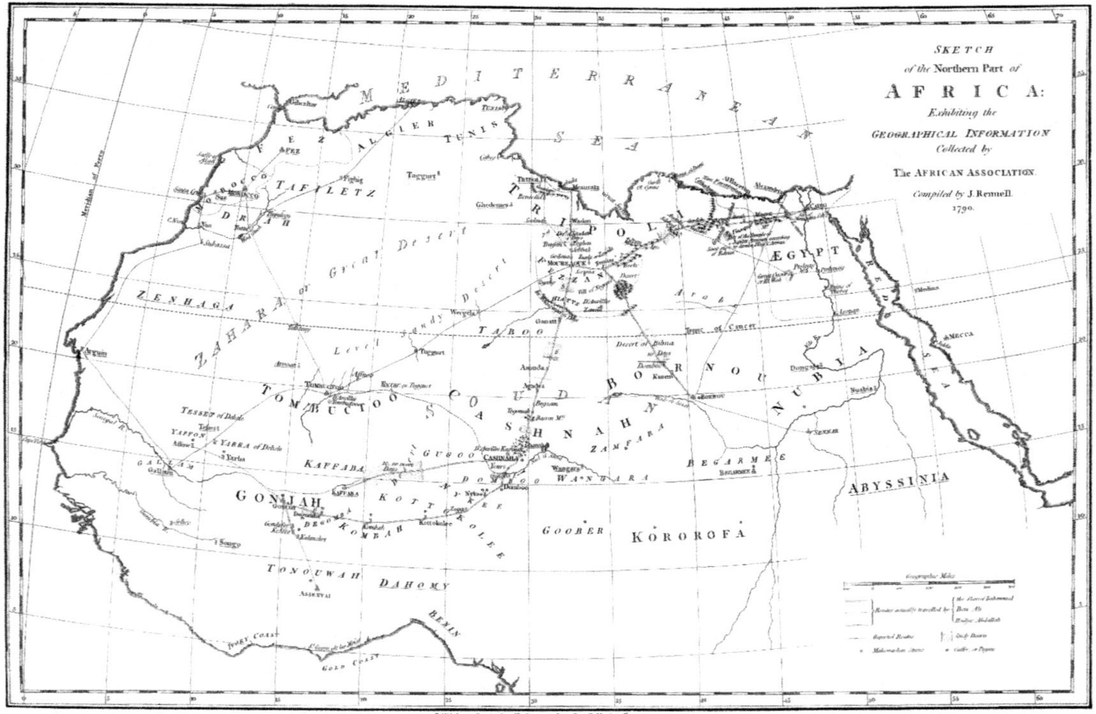

Published according to Act of Parliament by James Rennell, March 27th 1790.

PROCEEDINGS

OF THE

ASSOCIATION

FOR

PROMOTING THE DISCOVERY

OF THE

INTERIOR PARTS OF AFRICA.

LONDON:

PRINTED FOR T. CADELL, IN THE STRAND.

1791.

LIST OF THE MEMBERS

OF THE

ASSOCIATION.

A

THE Counteſs of Aileſbury
Right Hon. H. Addington, Speaker of the Houſe of Commons

B

Duke of Buccleugh
Earl of Buckinghamſhire
Earl of Bute
Earl of Briſtol
Sir Joſeph Banks, Bart. P. R. S.
Lady Belmore
Robert Barclay, Eſq,
Henry Beaufoy, Eſq. M. P.

Mark

Mark Beaufoy, Efq.
John Beaufoy, Efq.
Richard Henry Bennet, Efq.
Ifaac H. Browne, Efq. M. P.
Robert Barclay, Efq. M. P.

C

Lord Carysfort
Earl of Cholmondeley
Sir H. G. Calthorpe, Bart. M. P.
General Conway
John Call, Efq. M. P.
Mrs. Child
Thomas Coutts, Efq.
John Campbell, Efq. M. P.

D

Lord Daer
Sir John Dick, Bart.
Wm. Drake, Jun. Efq. M. P.

Earl

E

Earl of Exeter

G. N. Edwards, Efq. M. P.

F

Earl of Fife

Sir A. Ferguffon, Bart. M. P.

Sir William Fordyce, Bart.

Colonel Fullerton, M. P.

G

Duke of Grafton

Earl of Gainfborough

Earl of Galloway

Edward Gibbon, Efq.

Dr. Gifborne

George Goftling, Efq.

George Gomm, Efq.

H

Earl of Huntingdon

Lord Hawke

Mr. Profeſſor Harwood.
Sir John Hort, Bart.
John Hunter, Eſq.
Henry Hoare, Eſq.
Charles Hoare, Eſq.
Henry H. Hoare, Eſq.

K

Whitſhed Keene, Eſq. M. P.

L

Lord Loughborough
Biſhop of Landaff
Wilfred Lawſon, Eſq.
Dr. Lettſom
William Ludlam, Eſq.

M

Earl of Moira
Lord Middleton
Sir Charles Middleton, Bart.
Sir William Muſgrave, Bart.

William

William Marſden, Eſq.
Rev. Dr. Marton
Paul Le Meſurier, Eſq. M. P.
Charles Miller, Eſq.
James Martin, Eſq. M. P.

N

Duke of Northumberland
Richard Neave, Eſq.
The Hon. F. North

P

John Peachy, Eſq. M. P.
W. Pulteney, Eſq. M. P.
Charles A. Pelham, Eſq. M. P.
W. M. Pitt, Eſq. M. P.
John Parke, Eſq.

R

Earl of Radnor
Lord of Rawdon
Lieut. Gen. Rainsford

S

Lord Sheffield
Sir John Sinclair, Bart. M. P.
Sir John Stepney, Bart.
John Stanley, Efq.
Mr. Stuart
Hugh Scott. Efq.
John Simmons, Efq.
William Smith, Efq. M. P.
Richard Stonehewer, Efq.
Hans Sloane, Efq. M. P.

T

Greaves Townley, Efq.
Robert Thornton, Efq. M. P.

V

Benjamin Vaughan, Efq.

W

Earl of Wycombe
Sir Godfrey Webfter, Bart.

William

William Watſon, Eſq.
Samuel Whitebread, Eſq. M. P.
William Wilberforce, Eſq. M. P.
William Winch, Eſq.
Joſiah Wedgewood, Eſq.
John Wilkinſon, Eſq.

Y

Philip York, Eſq. M. P.

CONTENTS.

ney

CHAP-

CHAP-

CHAPTER X.

CHAPTER XI.

CHAPTER. XII.

INTRODUCTION.

THE Narrative of the Proceedings of the SOCIETY that was formed in the year 1788, for the purpofe of Promoting the Difcovery of the Inland Diftricts of Africa, was written, at the requeft of his Colleagues, by one of the Members of the Committee of that ASSOCIATION; and is now printed at the defire, and for the ufe of the SOCIETY. But as it may alfo be read by perfons unacquainted with the Origin and Object of the Undertaking to which it relates, the following Paper, as defcriptive of both, is republifhed for their information.

PLAN

OF THE

ASSOCIATION.

OF the objects of inquiry which engage our attention the moſt, there are none, perhaps, that ſo much excite continued curioſity, from childhood to age; none that the learned and unlearned ſo equally wiſh to inveſtigate, as the nature and hiſtory of thoſe parts of the world, which have not, to our knowledge, been hitherto explored. To this deſire the Voyages of the late Captain Cook have ſo far afforded gratification, that nothing worthy of reſearch by Sea, the Poles themſelves excepted, remains to be examined. But by Land, the objects of Diſcovery are ſtill ſo vaſt, as to include at leaſt a third of the habitable ſurface of the earth: for much of Aſia, a ſtill larger proportion of America, and almoſt the whole of Africa, are unviſited and unknown.

In Aſia there are few extenſive diſtricts of which we are wholly ignorant; but there are many of which we are imperfectly informed; and to our knowledge of ſeveral of theſe, the expected publication of the Travels of Mr. Foſter, in the ſervice of the Eaſt India Company, may bring material improvement. For, about three years ſince, in returning from Hindoſtan to Europe, he travelled by the way of Laldong, Jummoo, Caſhmire, Cabul, Herat, and the Caſpian Sea; and though the character of a Mooriſh Merchant,

chant, a difguife which the nature of the journey compelled him to affume, would not permit him to depart fo far from the ufage of Afia, as to make a draught of the country, or to write any other than fhort memorandums as he paffed, yet, if we may judge from the opportunities he had of information, his Narrative muft be important. It will probably fhew the manners and cuftoms, and military ftrength of the populous tribes that inhabit the mountains on the North of Lahore: it promifes to gratify the eagernefs which all men exprefs to acquire a knowledge of the fequeftered and unexplored, though celebrated Country of Cafhmire: and there is reafon to fuppofe, that it will alfo defcribe the rifing Empire of the Seiks, the conquerors of Zabeta Cawn, and the rivals of Abdalla. Should this be the cafe, we fhall learn the hiftory of an Empire that already extends from the river Attok, the weſtern branch of the Indus, to the banks of the Jumma; and poſſibly too we may alſo be told the particulars of a Religion, which, according to the accounts received, profeſſes to bring back the Hindoos from the idolatrous veneration of images to the purity of their primitive faith, the worſhip of One God: a Religion, which is ſaid to aſcribe to its Founder, Nanock, who died about 200 years ſince, a ſacred character, by ſuppoſing that he was Brimha, and that this was his laſt appearance upon earth: a Religion, which its Followers, in contradiction to the former uniform practice of the Believers in the Shaſter, endeavour to make univerſal, and with a zeal which reſembles the Mahometan, conſtantly enforce by the ſword.

To

To our knowledge of America, a large and valuable addition may ſoon be expected; for ſeveral of the inhabitants of Canada had the ſpirit, about two years ſince, to ſend, at their own expence, different perſons to traverſe that vaſt continent, from the river St. Lawrence weſtward to the oppoſite ocean.

While, in this manner, the circle of our knowledge with reſpect to Aſia and America is gradually extending itſelf, and advancing towards perfection, ſome progreſs has been made in the diſcovery of particular parts of Africa: for Dr. Sparrman's Narrative has furniſhed important information, to which will ſoon be added that of Mr. Patterſon, whoſe account of his Travels and Obſervations in the Southern Parts of Africa is already in the Preſs; and if a deſcription of the ſtill more extended Travels of Colonel Gordon, the preſent Commander of the Dutch Troops at the Cape of Good Hope, ſhould be given to the Public, the ſouthern extremity of the African Peninſula may perhaps be juſtly conſidered as explored. Mr. Bruce alſo, it is ſaid, is preparing for the Preſs an account of the knowledge which he has obtained on the eaſtern ſide of that quarter of the globe.*

But notwithſtanding the progreſs of diſcovery on the coaſts and borders of that vaſt continent, the map of its Interior is ſtill but a wide extended blank, on which the Geographer, on the authority of Leo Africanus, and of the Xeriff Edriſſi the Nubian Author, has traced, with a heſitating hand, a few names of unexplored rivers and of uncertain nations.

The

* Mr. Patterson's and Mr. Bruce's Travels are now publiſhed.

The courſe of the Niger, the places of its riſe and termination, and even its exiſtence as a ſeparate ſtream, are ſtill undetermined. Nor has our knowledge of the Senegal and Gambia rivers improved upon that of De la Brue and Moore; for though ſince their time half a century has elapſed, the Falls of Felu on the firſt of theſe two rivers, and thoſe of Baraconda on the laſt, are ſtill the limits of diſcovery.

Neither have we profited by the information which we have long poſſeſſed, that even on the weſtern coaſts of Africa, the Mahometan faith is received in many extenſive diſtricts, from the Tropic of Cancer ſouthward to the Line. That the Arabic, which the Muſſelman Prieſts of all countries underſtand, furniſhes an eaſy acceſs to ſuch knowledge as the weſtern Africans are able to ſupply, is perfectly obvious; as it alſo is, that thoſe

Africans

Africans muſt, from the nature of their Religion, poſſeſs, what the Traders to the coaſt aſcribe to them, an intercourſe with Mecca. But although theſe circumſtances apparently prove the practicability of exploring the Interior Parts of Africa, and would much facilitate the execution of the Plan, yet no ſuch efforts have hitherto been made. Certain however it is, that, while we continue ignorant of ſo large a portion of the globe, that ignorance muſt be conſidered as a degree of reproach upon the preſent age.

Senſible of this ſtigma, and deſirous of reſcuing the age from a charge of ignorance, which, in other reſpects, belongs ſo little to its character, a few Individuals, ſtrongly impreſſed with a conviction of the practicability and utility of thus enlarging the fund of human knowledge,

knowledge, have formed the Plan of an Aſſociation for Promoting the Diſcovery of the Interior Parts of Africa.

The nature of their Eſtabliſhment will beſt appear from the following account of their proceedings.

At an ADJOURNED MEETING of the SATURDAY'S CLUB, at the *St. Alban's Tavern*, on the 9th of *June*, 1788,

PRESENT,

EARL OF GALLOWAY,
LORD RAWDON,
GENERAL CONWAY,
SIR ADAM FERGUSSON,
SIR JOSEPH BANKS,
SIR WILLIAM FORDYCE,
MR. PULTNEY,
MR. BEAUFOY,
MR. STUART:

ABSENT

ABSENT MEMBERS.

BISHOP OF LANDAFF,
LORD CARYSFORT,
SIR JOHN SINCLAIR.

RESOLVED,

That as no ſpecies of information is more ardently deſired, or more generally uſeful, than that which improves the ſcience of Geography; and as the vaſt Continent of Africa, notwithſtanding the efforts of the Antients, and the wiſhes of the Moderns, is ſtill in a great meaſure unexplored, the Members of this Club do form themſelves into an Aſſociation for Promoting the Diſcovery of the Inland Parts of that Quarter of the World:

That, for the ſaid purpoſe, each Member do ſubſcribe Five Guineas a year, for three years; and that at, or after that period,

period, any Member, on giving a year's notice, may withdraw himſelf from the Aſſociation:

That during the firſt twelve months from the preſent day, each of the Members of the Club be allowed to recommend, for the approbation of the Club, ſuch of his Friends as he ſhall think proper to be admitted to the new Aſſociation; but that after that time all additional Members be elected by a Ballot of the Aſſociation at large:

That a Committee, conſiſting of a Secretary, Treaſurer, and three Aſſiſting Members, be choſen by Ballot:

That the ſaid Committee do prepare and ſubmit to the conſideration of the Members, at their next meeting, ſuch Rules as they ſhall think requiſite for the

the effectual attainment of the object of the new Inſtitution, and for its good government:

That the Committee be entruſted with the choice of the perſons who are to be ſent on the Diſcovery of the Interior Parts of Africa, together with the Society's Correſpondence, and the Management of its Funds:

That the Committee ſhall not diſcloſe, except to the Members of the Aſſociation at large, ſuch intelligence as they ſhall, from time to time, receive from the perſons who ſhall be ſent out on the buſineſs of Diſcovery:

That on the receipt of any intereſting intelligence from any of the ſaid perſons, the Members of the Aſſociation ſhall be convened by Letters from the

Secretary;

Secretary; and that ſuch parts of the ſaid intelligence as, in the opinion of the Committee, may, without endangering the object of their Aſſociation, be made public, ſhall be communicated to the Meeting:

That an Account of all Monies paid and received ſhall, on the laſt Saturday in the month of May in each year, be ſubmitted to the conſideration of the Society at large, by the Treaſurer:

That the Members of the Committee be choſen by Ballot, on the firſt Saturday in the month of May in each year.

The preceding Reſolutions having been agreed to by all the Members preſent, they proceeded on the ſame day, the

the 9th of June, 1788, in purſuance of their Fourth Reſolution, to chuſe a Committee by Ballot, and the following perſons were elected:

LORD RAWDON,
BISHOP OF LANDAFF,
SIR JOSEPH BANKS,
MR. BEAUFOY,
MR. STUART.

CHAPTER I.

Proceedings of the Aſſociation, from the Time of its Eſtabliſhment, to that of the Departure of Mr. LEDYARD.

THE Aſſociation for Promoting the Diſcovery of the Interior Regions of Africa was formed on the 9th of June, in the year 1788; and on the ſame day a Committee of its Members was inveſted with the Direction of its Funds, the Management of its Correſpondence, and the Choice of the Perſons to whom the Geographical Miſſion ſhould be aſſigned.

Naturally anxious for the ſpeedy attainment of the important object thus recommended to their care, an object made doubly interesting by the consideration of its having engaged the attention, and baffled the researches of the most inquisitive and most powerful nations of antiquity, the Managers proceeded with the utmost ardour to the immediate execution of the Plan.

Two Gentlemen, whose qualifications appeared to be eminent, proposed to undertake the Adventure.

One of them, a Mr. LEDYARD, was an American by birth, and seemed from his youth to have felt an invincible desire to make himself acquainted with the unknown, or imperfectly discovered regions of the globe. For several years he had lived with the Indians of America, had studied their manners, and had practised in their school the means of

obtaining

obtaining the protection, and of recommending himself to the favour of Savages. In the humble situation of a Corporal of Marines, to which he submitted rather than relinquish his pursuit, he had made, with Captain Cook, the Voyage of the World; and feeling on his return an anxious desire of penetrating from the North Western Coast of America, which Cook had partly explored, to the Eastern Coast, with which he himself was perfectly familiar, he determined to traverse the vast Continent from the Pacific to the Atlantic Ocean.

His first Plan for the purpose was that of embarking in a vessel which was then preparing to sail, on a Voyage of Commercial Adventure, to Nootka Sound, on the Western Coast of America; and with this view he expended in sea stores, the greateſt part of the money which his chief benefactor Sir Joſeph Banks (whoſe generous conduct the Writer of this Narrative has often heard him acknowledge) had liberally ſupplied. But the ſcheme being fruſtrated by the rapacity of a Cuſtom-houſe Officer, who had ſeized and detained the veſſel for reaſons which on legal inquiry proved to be frivolous, he determined to travel over land to Kamſchatka, from whence, to the Weſtern Coaſt of America, the paſſage is extremely ſhort. With no more than ten guineas in his purſe, which was all that he had left, he croſſed the Britiſh Channel to Oſtend, and by the way of Denmark and the Sound, proceeded to the capital of Sweden, from which, as it was Winter, he attempted to traverſe the Gulph of Bothnia on the ice, in order to reach Kamſchatka by the ſhorteſt way; but finding, when he came to the middle

middle of the ſea, that the water was not frozen, he returned to Stockholm, and taking his courſe Northward, walked into the Arctic Circle; and paſſing round the head of the Gulph, deſcended on its Eaſtern ſide to Peterſburgh.

There he was ſoon noticed as an extraordinary man. Without ſtockings, or ſhoes, and in too much poverty to provide himſelf with either, he received and accepted an invitation to dine with the Portugueze Ambaſſador. To this invitation it was probably owing that he was able to obtain the ſum of twenty guineas for a bill on Sir Joſeph Banks, which he confeſſed he had no authority to draw, but which, in conſideration of the buſineſs that he had undertaken, and of the progreſs that he had made, Sir Joſeph, he believed, would not be unwilling to pay. To the Ambaſſador's intereſt it might alſo be owing that he obtained permiſſion to accompany a detachment of Stores which the Empreſs had ordered to be ſent to Yakutz, for the uſe of Mr. Billings, an Engliſhman, at that time in her ſervice.

Thus accommodated, he travelled Eaſtward through Siberia, ſix thouſand miles, to Yakutz, where he was kindly received by Mr. Billings, whom he remembered on board Captain Cook's ſhip, in the ſituation of the Aſtronomer's Servant, but to whom the Empreſs had now entruſted her ſchemes of Northern Diſcovery.

From Yakutz he proceeded to Oczakow, on the coaſt of the Kamſchatka ſea, from whence he meant to have paſſed over to that peninſula, and to have embarked on the Eaſtern ſide in one of the Ruſſian veſſels that trade to the Weſtern ſhores

ſhores of America; but finding that the navigation was completely obſtructed by the ice, he returned again to Yakutz, in order to wait for the concluſion of the Winter.

Such was his ſituation when, in conſequence of ſuſpicions not hitherto explained, or reſentments for which no reaſon is aſſigned, he was ſeized, in the Empreſs's name, by two Ruſſian ſoldiers, who placed him in a ſledge, and conveying him, in the depth of Winter, through the Deſarts of the Northern Tartary, left him at laſt on the Frontiers of the Poliſh Dominions. As they parted they told him, that if he returned to Ruſſia, he would certainly be hanged; but that if he choſe to go back to England, they wiſhed him a pleaſant journey.

In the midſt of poverty, covered with rags, infeſted with the uſual accompaniments of ſuch cloathing, worn with continued hardſhip, exhauſted by diſeaſe, without friends, without credit, unknown, and full of miſery, he found his way to Koningſberg.---There, in the hour of his uttermoſt diſtreſs, he reſolved once more to have recourſe to his old Benefactor, and he luckily found a perſon who was willing to take his draft for five guineas on the Preſident of the Royal Society.

With this aſſiſtance he arrived in England, and immediately waited on Sir Joſeph Banks, who told him, knowing his temper, that he believed he could recommend him to an adventure almoſt as perilous as the one from which he had returned; and then communicated to him the wiſhes of the Aſſociation for

Diſ-

Difcovering the Inland Countries of Africa.

Ledyard replied, that he had always determined to traverfe the Continent of Africa as foon as he had explored the Interior of North America; and as Sir Jofeph had offered him a Letter of Introduction, he came directly to the Writer of thefe Memoirs. Before I had learnt from the note the name and bufinefs of my Vifitor, I was ftruck with the manlinefs of his perfon, the breadth of his cheft, the opennefs of his countenance, and the inquietude of his eye. I fpread the map of Africa before him, and tracing a line from Cairo to Sennar, and from thence Weftward in the latitude and fuppofed direction of the Niger, I told him that was the route, by which I was anxious that Africa might, if poffible, be explored. He faid, he

fhould

ſhould think himſelf ſingularly fortunate to be entruſted with the Adventure. I aſked him when he would ſet out? "To-morrow morning," was his anſwer. I told him I was afraid that we ſhould not be able, in ſo ſhort a time, to prepare his inſtructions, and to procure for him the letters that were requiſite; but that if the Committee ſhould approve of his propoſal, all expedition ſhould be uſed.

Such is the hiſtory, and ſuch were the qualifications of one of the perſons whom the Committee engaged in its ſervice.

The other, Mr. LUCAS, had been ſent, when a boy, to Cadiz, in Spain, for education as a merchant, and having the misfortune, on his return, to be captured by a Sallee Rover, was brought as a ſlave a ſlave to the Imperial Court of Morocco.

Three years of captivity preceded his reſtoration to freedom, and his conſequent departure for Gibraltar; where, at the requeſt of General Cornwallis, he accepted the offices of Vice-Conſul and Chargé d'Affaires in the Empire of Morocco; and had the ſatisfaction to return, as the Delegate of his Sovereign, to the very kingdom in which, for ſo long a period, he had lived as a ſlave. At the end of ſixteen years he once more reviſited England, and was ſoon appointed Oriental Interpreter to the Britiſh Court, in which ſituation he was when he became known to the Committee, and expreſſed his willingneſs, with His Majeſty's permiſſion, to undertake, in the Service of the Aſſociation, whatever Journey his knowledge of the

the Manners, Cuſtoms, and Language of the Arabs might enable him to perform. His Majeſty, with that liberal attention to the Progreſs of Knowledge which at all times has diſtinguiſhed his reign, ſignified his pleaſure, that Mr. LUCAS ſhould proceed on the buſineſs of the Society; and that his ſalary, as Oriental Interpreter, ſhould be continued to him during his abſence.

Having thus obtained the aſſiſtance of two perſons ſo eminently qualified to facilitate the attainment of its object, the Committee proceeded to preſcribe to them their reſpective routes.

To Mr. LEDYARD they aſſigned, at his own deſire, as an enterprize of obvious peril and of difficult ſucceſs, the taſk of traverſing from Eaſt to Weſt, in the latitude attributed to the Niger,

the

the wideſt part of the Continent of Africa.

To Mr. LUCAS, in conſideration of the knowledge which he poſſeſſed of the Language and Manners of the Arabs, they allotted the paſſage of the Deſart of Zahara, from Tripoli to Fezzan; for they had learned from various information, that with this kingdom, which in ſome meaſure is dependent on Tripoli, the traders of Agadez and Tombuctou, and of other towns in the Interior of Africa, had eſtabliſhed a frequent and regular intercourſe; and their inſtructions to him were, that he ſhould proceed directly to Fezzan; that he ſhould collect and tranſmit by the way of Tripoli, whatever intelligence, reſpecting the Inland Regions of the Continent, the people of Fezzan, or the traders who viſited vifited their country, might be able to afford; and that he fhould afterwards return by the way of the Gambia, or by that of the Coaft of Guinea.

One obftacle to the departure of thefe Geographical Miffionaries was ftill to be removed; and that was, the fmallnefs of the Fund; for the Members of the Affociation, which had not yet paffed the fecond month of its exiftence, were extremely few, and the Committee were too confcious of the importance and dignity of their undertaking, to canvafs for fubfcriptions.

In this dilemma, the Committee refolved to advance the money that was requifite; and they accordingly raifed among themfelves the fum of 430l. which enabled them to provide for their travellers

vellers the means of immediate equipment, and the letters of neceſſary credit.

Mr. LUCAS, having been detained by illneſs, did not leave England till the 6th of Auguſt.

CHAPTER II.

Mr. LEDYARD'S *Arrival at Cairo.—His Remarks on the Inhabitants, &c.—His Death and Character.*

MR. LEDYARD took his departure from London on the 30th of June, 1788; and after a journey of ſix and thirty days, ſeven of which were conſumed at Paris, and two at Marſeilles, arrived in the city of Alexandria.

His Letters of Recommendation to the Britiſh Conſul ſecured him from the embarraſſments which the want of inns

would

would otherwiſe have occaſioned; and procured for him the neceſſary inſtructions for aſſuming the dreſs, and adopting the manners that are requiſite for an Egyptian Traveller.

Forcibly impreſſed by the objects which he ſaw, and naturally led to compare them with thoſe which other Regions of the Globe had preſented to his view, he deſcribes with the energy of an original Obſerver, and exhibits in his Narrative the varied effect of ſimilarity and contraſt. But as the travellers who preceded him, have obtained and tranſmitted to Europe whatever knowledge, either antient or modern, the Lower Egypt affords, and as the examination of that country was no part of the buſineſs which was given him in charge, his deſcriptions, generally ſpeaking, would add but

but little to the inſtruction which other Narratives convey.

The following Extracts, from different parts of his Journal, are given in his own words.

" A traveller, who ſhould, by juſt compariſons between things here and in Europe, tell his tale; who, by a mind unbewitched by antecedent deſcriptions, too ſtrong, too bold, too determined, too honeſt, to be capable of lying, ſhould ſpeak juſt as he thought, would, no doubt, be eſteemed an arrant fool, and a ſtupid coxcomb.—For example, an Engliſhman who had never ſeen Egypt, would aſk me what ſort of a woman an Egyptian woman was? If I meant to do the queſtion as much juſtice by the anſwer, as I could in my way, I ſhould aſk him to take notice of the firft company of Gypfies he faw behind a hedge in Effex; and I fuppofe he would be fool enough to think me a fool.

"August 14. I left Alexandria at midnight, with a pleafant breeze at North; and was, at fun-rife next morning, at the mouth of the Nile, which has a bar of fand acrofs it, and foundings as irregular as the fea, which is raifed upon it by the contentions of counter currents and winds.

"The view in failing up the Nile is very confined, unlefs from the top of the maft, or fome other eminence, and then it is an unbounded plain of excellent land, miferably cultivated, and yet interfperfed with a great number of villages, both on its banks and as far along

along the meadows as one can ſee in any direction. The river is alſo filled with boats paſſing and repaſſing----boats all of one kind, and navigated in one manner; nearly alſo of one ſize, the largeſt carrying ten or fifteen tons. On board of theſe boats are ſeen onions, water-melons, dates, ſometimes a horſe, a camel, (which lies down in the boat) and ſheep and goats, dogs, men and women.---Towards evening and morning they have muſic.

" Whenever we ſtopped at a village, I uſed to walk into it with my Conductor, who, being a Muſſelman, and a deſcendant from Mahommed, wore a green turban, and was therefore reſpected, and I was ſure of ſafety:—but in truth, dreſſed as I was in a common Turkiſh habit, I believe I ſhould have walked as ſafely without him. I ſaw no propenſity among the inhabitants to incivility. The villages are moſt miſerable aſſemblages of poor little mud huts, flung very cloſe together without any kind of order, full of duſt, lice, fleas, bed-bugs, flies, and all the curſes of Moſes; people poorly clad, the youths naked: in ſuch reſpect, they rank infinitely below any Savages I ever ſaw.

"The common people wear nothing but a ſhirt and drawers, and they are always blue. Green is the royal or holy colour; none but the deſcendants of Mahommed, if I am rightly informed, being permitted to wear it.

"August 19. From the little town where we landed, the diſtance to Cairo is about a mile and a half, which we rode on aſſes; for the aſs in this country is the Chriſtian's horſe, as he is allowed no no other animal to ride upon. Indeed I find the ſituation of a Chriſtian, or what they more commonly call here a Frank, to be very, very humiliating, ignominious, and diſtreſſing. No one, by a combination of any cauſes, can reaſon down to ſuch effects as experience teaches us do exiſt here; it being impoſſible to conceive, that the enmity I have alluded to could exiſt between men;——or, in fact, that the ſame ſpecies of beings, from any cauſes whatever, ſhould ever think and act ſo differently as the Egyptians and the Engliſh do.

"I arrived at Cairo early in the morning, on the 19th of Auguſt, and went to the houſe of the Venetian Conſul, Mr. Roſetti, Chargé d'Affaires for the Engliſh Conſul here.

"After dinner, not being able to find

any other lodging, and receiving no very preſſing invitation from Mr. Roſetti to lodge with him, I went to a convent. This convent conſiſts of Miſſionaries ſent by the Pope to propagate the Chriſtian Faith, or at leaſt to give ſhelter to Chriſtians. The Chriſtians here are principally from Damaſcus: the convent is governed by the Order of Recollets: a number of Engliſh, as well as other European travellers, have lodged there.

" Auguſt 21ſt. It is now about the hotteſt ſeaſon of the year here; but I think I have felt it warmer in the City of Philadelphia, in the ſame month.

" Auguſt 26th. This day I was introduced by Roſetti to the Aga Mahommed, the confidential Miniſter of Iſmael, the moſt powerful of the four ruling Beys: he gave me his hand to kiſs, and

with

with it the promiſe of letters, protection, and ſupport, through Turkiſh Nubia, and alſo to ſome Chiefs far inland. In a ſubſequent converſation, he told me I ſhould ſee in my travels a people who had power to tranſmute themſelves into the forms of different animals. He aſked me what I thought of the affair? I did not like to render the ignorance, ſimplicity, and credulity of the Turk apparent. I told him, that it formed a part of the character of all Savages to be great Necromancers; but that I had never before heard of any ſo great as thoſe which he had done me the honour to deſcribe; that it had rendered me more anxious to be on my voyage; and if I paſſed among them, I would, in the letter I promiſed to write to him, give him a more particular account of them than he had hitherto had. ---He aſked me how I could travel without the language of the people where I ſhould

ſhould paſs? I told him, with vocabularies:---I might as well have read to him a page of Newton's Principia. He returned to his fables again. Is it not curious, that the Egyptians (for I ſpeak of the natives of the country as well as of him, when I make the obſervation) are ſtill ſuch dupes to the arts of ſorcery? Was it the ſame people that built the Pyramids?

" I can't underſtand that the Turks have a better opinion of our mental powers than we have of theirs; but they ſay of us, that we are " *a people who carry our minds on our fingers ends:*" meaning, that we put them in exerciſe conſtantly, and render them ſubſervient to all manner of purpoſes, and with celerity, diſpatch, and eaſe, do what we do.

" I ſuſpect the Copts to have been the origin

origin of the Negro race: the nofe and lip correfpond with thofe of the Negro. The hair, whenever I can fee it among the people here, (the Copts) is curled; —not clofe like the Negros, but like the Mulattoes. I obferve a greater variety of colour among the human fpecies here than in any other country; and a greater variety of feature than in any other country not poffeffing a greater degree of civilization.

" I have feen an Abyffinian woman and a Bengal man------the colour is the fame in both; fo are their features and perfons.

" I have feen a fmall mummy;----it has what I call wampum work on it. It appears as common here as among the Tartars. Tatowing is as prevalent among the Arabs of this place as among the

the South Sea Iſlanders. It is a little curious, that the women here are more generally than in any other part of the world tatowed on the chin, with perpendicular lines deſcending from the under lip to the chin, like the women on the North Weſt Coaſt of America. It is alſo a cuſtom here to ſtain the nails red, like the Cochin Chineſe, and the Northern Tartars. The maſk or veil that the women here wear, reſembles exactly that worn by the Prieſts at Otaheite, and thoſe ſeen at Sandwich Iſlands.

"I have not yet ſeen the Arabs make uſe of a tool like our axe or hatchet; but what they uſe for ſuch purpoſes as we do our hatchet and axe, is in the form of an adze, and is a form we found moſt agreeable to the South Sea Iſlanders. I ſee no inſtance of a tool formed deſignedly for the uſe of the right or left

left hand particularly, as the cotogon is among the Yorkertic Tartars.

" There is certainly a very remarkable affinity between the Ruſſian and Greek dreſs. The fillet round the temples of the Greek and Ruſſian women, is a circumſtance in dreſs that perhaps would ſtrike nobody as it does me; and ſo of the wampum work too, which is alſo found among them both.

" They ſpin here with the diſtaff and ſpindle only, like the French peaſantry and others in Europe; and the common Arab loom is upon our principle, though rude.

" I ſaw to-day (Auguſt 10th) an Arab woman white, like the White Indians in the South Sea Iſlands, Iſtmus of Darien, &c. Theſe kind of people all look alike.

" Among

"Among the Greek women here, I find the identical Archangel head-drefs.

"Their mufic is inftrumental, confifting of a drum and pipe, both which refemble thofe two inftruments in the South Seas. The drum is exactly like the Otaheite drum; the pipe is made of cane, and confifts of a long and fhort tube joined: the mufic refembles very much the bagpipe, and is pleafant.---All their mufic is concluded, if not accompanied, by the clapping of hands. I think it fingular, that the women here make a noife with their mouths like frogs, and that this frog-mufic is always made at weddings; and I believe on all other occafions of merriment where there are women.

"It is remarkable, that the dogs here are

are of juſt the ſame ſpecies found among the Otaheiteans.

" It is alſo remarkable, that in one village I ſaw exactly the ſame machines uſed for diverſion as in Ruſſia.---I forget the Ruſſian name for it. It is a large kind of wheel, on the extremities of which there are ſuſpended ſeats, in which people are whirled round over and under each other.

" The women dreſs their hair behind exactly in the ſame manner in which the women of the Calmuc Tartars dreſs theirs.

" In the Hiſtory of the Kingdom of Benin in Guinea, the Chiefs are called Aree Roee, or Street Kings. Among the the Iſlands in the South Sea, Otaheite, &c. they call the Chiefs Arees; and the great

great Chiefs, Aree le Hoi. I think this curious; and ſo I do that it is a cuſtom among the Arabs to ſpread a blanket when they would invite any one to tea or reſt with them.—American Indians ſpread the beaver ſkins on ſuch occaſions.

" The Arabs of the Deſarts, like the Tartars, have an invincible attachment to Liberty. No arts will reconcile them to any other life, or form of government, however modified. This is a character given me here of the Arabs.

" It is ſingular that the Arab Language has no word for Liberty, although it has for Slaves.

" The Arabs, like the New Zealanders, engage with a long ſtrong ſpear.

" I have made the beſt enquiries I have

have been able, ſince I have been here, of the nature of the country before me; of Sennar, Darfoor, Wangara, of Nubia, Abyſſinia, of thoſe named, or unknown by name. I ſhould have been happy to have ſent you better information of thoſe places than I am yet able to do. It will appear very ſingular to you in England, that we in Egypt are ſo ignorant of countries which we annually viſit. The Egyptians know as little of Geography as the generality of the French; and like them, ſing, dance, and traffic without it.

" I have the beſt aſſurances of a certain and ſafe conduct by the return of the caravan that is arrived from Sennar; and Mr. Roſetti tells me that the letters I ſhall have from the Aga here, will inſure me of being conveyed, from hand to hand, to my journey's end.

" The Mahometans in Africa are what the Ruſſians are in Siberia—a trading, enterprizing, ſuperſtitious, warlike ſet of vagabonds, and wherever they are ſet upon going, they will and do go; but they neither can nor do make voyages merely commercial, or merely religious, acroſs Africa; and where we do not find them in commerce, we find them not at all. They cannot (however vehemently puſhed on by religion) afford to croſs the Continent without trading by the way.

" October 14th. I went to-day to the market-place, where they vend the Black ſlaves that come from towards the interior parts of Africa:---there were 200 of them together, dreſſed and ornamented as in their own country. The appearance of a Savage in every region is almoſt the ſame!—There were very few men

men among them: this indicates that they are prifoners of war. They have a great many beads and other ornaments about them that are from the Eaft. I was told by one of them that they came from the Weft of Sennar, fifty-five days journey, which may be about four or five hundred miles. A Negro Chief faid, the Nile had its fource in his country. In general they had their hair plaited in a great number of fmall detached plaits, none exceeding in length fix or eight inches—the hair was filled with greafe, and dirt purpofely daubed on.

" October 16th. I have renewed my vifit to-day, and paffed it more agreeably than yefterday; for yefterday I was rudely treated. The Franks are prohibited to purchafe flaves, and therefore the Turks do not like to fee them in the market. Mr. Rofetti favoured me one of his running Chargé d'Affaires to accompany me: but having obferved yefterday among the ornaments of the Negros a variety of beads, and wanting to know from what country they came, I requefted Mr. Rofetti, previoufly to my fecond vifit, to fhew me from his ftore famples of Venetian beads.——He fhewed me famples of fifteen hundred different kinds. After this I fet out.

"The name of the country thefe Savages come from is Darfoor, and is well known on account of the Slave Trade, as well as of that in Gum and Elephant's teeth.

"The appearance of thefe Negros declares them to be a people in as favage a ftate as any people can; but not of fo favage a temper, or of that fpecies of countenance that indicates favage intelligence.

gence. They appear a harmlefs wild people; but they are moftly young women.

" The beads they are ornamented with are Venetian; and they have fome Venetian brafs medals which the Venetians make for trade. The beads are worked wampum-wife. I know not where they got the marine fhells they worked among their beads, nor how they could have feen white men. I afked them if they would ufe me well in their country, if I fhould vifit it? They faid, " Yes:"— and added, that they fhould make a King of me, and treat me with all the delicacies of their country. Like the Egyptian women, and like moft other Savages, they ftick on ornaments wherever they can, and wear, like them, a great ring in the nofe, either from the cartilage, or from the fide: they alfo rub on fome black kind of paint round the eyes,

 like

like the Egyptian women. They are a ſizeable well-formed people, quite black, with what, I believe, we call the true Guinea face, and with curled ſhort hair; but not more curled or ſhorter than I have ſeen it among the Egyptians; but in general theſe Savages plait it in taſſels plaiſtered with clay and paint. Among ſome of them the hair is a foot long, and curled, reſembling exactly one of our mops. The prevailing colour, where it can be ſeen, is a black and red mixed. I think it would make any hair curl, even Uncle Toby's wig, to be plaited and plaiſtered as this is. This caravan, which I call the Darfoor caravan, is not very rich.——The Sennar is the rich caravan.

"October 19th. I went yeſterday to ſee if more of the Darfoor caravan had arrived; but they were not. I wonder why

why travellers to Cairo have not visited these slave markets, and conversed with the Jelabs or travelling Merchants of these caravans: both are certainly sources of great information.—The eighth part of the money expended on other accounts, might here answer some good solid purpose. For my part, I have not expended a crown, and I have a better idea of the people of Africa, of its trade, of the position of places, the nature of the country, manner of travelling, &c. than ever I had by any other means; and, I believe, better than any other means would afford me.

"October 25th. I have been again to the slave market; but neither the Jelabs (a name which in this country is given to all travelling Merchants) nor the slaves are yet arrived in town—— they will be here to-morrow. I met two or three in the ſtreet, and one with a ſhield and ſpear.

"I have underſtood to-day, that the King of Sennar is himſelf a Merchant, and concerned in the Sennar caravans. The Merchant here who contracts to convey me to Sennar, is Procurer at Cairo to the King of Sennar: this is a good circumſtance, and one I knew not of till to-day. Mr. Roſetti informed me of it. He informed me alſo, that the importation of Negro ſlaves into Egypt will amount to 20,000.—The caravans from the interior countries of Africa do not arrive here uniformly every year—they are ſometimes abſent two or three years.

"Among a dozen of Sennar ſlaves, I ſaw three perſonable men, of a good bright olive colour, of vivacious and intelligent

telligent countenances; but they had all three (which firſt attracted my notice) heads uncommonly formed: the forehead was the narroweſt, the longeſt, and moſt protuberant I ever ſaw. Many of theſe ſlaves ſpeak a few words of the Arab language; but whether they learned them before or ſince their captivity I cannot tell.

" A caravan goes from here (Cairo) to Fezzan, which they call a journey of fifty days; and from Fezzan to Tombuctou, which they call a journey of ninety days. The caravans travel about twenty miles a day, which makes the diſtance on the road from here to Fezzan, one thouſand miles; and from Fezzan to Tombuctou, one thouſand eight hundred miles. From here to Sennar is reckoned ſix hundred miles.

" I have

" I have been waiting ſeveral days to have an interview with the Jelabs who go from hence to Sennar. I am told that they carry, in general, trinkets; but among other things, ſoap, antimony, red linen, razors, ſciſſars, mirrors, beads; and, as far as I can yet learn, they bring from Sennar elephant's teeth, the gum called here Gum Sennar, camels, oſtrich feathers, and ſlaves.

" Wangara is talked of here as a place producing much gold, and as a kingdom: all accounts, and there are many, agree in this. The King of Wangara (whom I hope to ſee in about three months after leaving this) is ſaid to diſpoſe of juſt what quantity he pleaſes of his gold—ſometimes a great deal, and ſometimes little or none; and this, it is ſaid, he does to prevent ſtrangers knowing

ing how rich he is, and that he may live in peace."

Such are the moſt material of thoſe remarks on the people of Africa, which Mr. LEDYARD was enabled, by his reſidence at Cairo, to ſend to the Committee.——The views which they opened were interesting and inſtructive; but they derived their principal importance from the proofs which they afforded of the ardent ſpirit of inquiry, the unwearied attention, the perſevering reſearch, and the laborious, indefatigable, anxious zeal with which their Author purſued the object of his Miſſion.

Already informed that his next diſpatch would be dated from Sennar; that letters of earneſt recommendation had been given him by the Aga; that the terms of his paſſage had been ſettled; and

and that the day of his departure was appointed——the Committee expected with impatience the defcription of his journey. Great was therefore their concern, and fevere their difappointment, when letters from Egypt announced to them the melancholy tidings of his death. A bilious complaint, the confequence of vexatious delays in the promifed departure of the caravan, had induced him to try the effect of too powerful a dofe of the acid of vitriol; and the fudden uneafinefs and burning pain which followed the incautious draught, impelled him to feek relief from the violent action of the ftrongeft Tartar emetic. A continued difcharge of blood difcovered the danger of his fituation, and fummoned to his aid the generous friendfhip of the Venetian Conful, and the ineffectual fkill of the the moft approved phyficians of Cairo.

He

He was decently interred in the neighbourhood of ſuch of the Engliſh as had ended their days in the capital of Egypt.

The bilious complaint with which he was ſeized has been attributed to the frowardneſs of a childiſh impatience.—Much more natural is the conjecture, that his unexpected detention, week after week, and month after month, at Cairo, (a detention which conſumed his finances, which therefore expoſed to additional hazard the ſucceſs of his favourite enterprize, and which conſequently tended to bring into queſtion his honour to the Society) had troubled his ſpirits, had preyed upon his peace, and ſubjected him at laſt to the diſeaſe that proved in its conſequences the means of dragging him to his grave.

Of his attachment to the Society, and of

of his zeal for their ſervice, the following Extracts from his Letters are remarkably expreſſive:

"Money! it is a vile ſlave!—I have at preſent an œconomy of a more exalted kind to obſerve. I have the eyes of ſome of the firſt men of the firſt kingdom on earth turned upon me. I am engaged by thoſe very men, in the moſt important object that any private individual can be engaged in: I have their approbation to acquire, or to loſe; and their eſteem alſo, which I prize beyond every thing, except the independent idea of ſerving mankind. Should raſhneſs or deſperation carry me through, whatever fame the vain and injudicious might beſtow, I ſhould not accept of it;—it is the good and great I look to: fame from them beſtowed is altogether different, and is cloſely allied to a well-done from God;

God: but raſhneſs will not be likely to carry me through, any more than timid caution. To find the neceſſary medium of conduct, to vary and apply it to contingencies, is the œconomy I allude to; and if I ſucceed by ſuch means, men of ſenſe in any ſucceeding epoch will not bluſh to follow me, and perfect thoſe Diſcoveries I have only abilities to trace out roughly, or, a diſpoſition to attempt.

"A Turkiſh ſopha has no charms for me: if it had, I could ſoon obtain one here. I could to-morrow take the command of the beſt armament of Iſhmael Bey.—I ſhould be ſure of ſucceſs, and its conſequential honours. Believe me, a ſingle well-done from your Aſſociation has more worth in it to me, than all the trappings of the Eaſt; and what is ſtill more precious, is, the pleaſure I have in the

the juſtification of my own conduct at the tribunal of my own heart."

To thoſe who have never ſeen Mr. LEDYARD, it may not, perhaps, be uninintereſting to know, that his perſon, though ſcarcely exceeding the middle ſize, was remarkably expreſſive of activity and ſtrength; and that his manners, though unpoliſhed, were neither uncivil nor unpleaſing. Little attentive to difference of rank, he ſeemed to conſider all men as his equals, and as ſuch he reſpected them. His genius, though uncultivated and irregular, was original and comprehenſive. Ardent in his wiſhes, yet calm in his deliberations; daring in his purpoſes, but guarded in his meaſures; impatient of controul, yet capable of ſtrong endurance; adventurous beyond the conception of ordinary men, yet wary and conſiderate; and attentive to all precautions,

cautions, he appeared to be formed by Nature for atchievements of hardihood and peril.

They who compare the extent of his pilgrimage through the vaſt regions of Tartary with the ſcantineſs of his funds, will naturally aſk, by what means he obtained a ſubſiſtence on the road? All that I have ever learned from him on the ſubject, was, that his ſufferings were exceſſive, and that more than once he owed his life to the compaſſionate temper of the women. This laſt remark is ſtrongly confirmed by the following Extract from his account of his Siberian Tour.

" I have always remarked, that women, in all countries, are civil, obliging, tender, and humane; that they are ever inclined to be gay and chearful, timorous and modeſt; and that they do not heſitate, like men, to perform a generous action.—Not haughty, not arrogant, not ſupercilious, they are full of courteſy, and fond of ſociety: more liable, in general, to err than man; but in general, alſo, more virtuous, and performing more good actions than he. To a woman, whether civilized or ſavage, I never addreſſed myſelf in the language of decency and friendſhip, without receiving a decent and friendly anſwer. With man it has often been otherwiſe.

" In wandering over the *barren* plains of *inhoſpitable Denmark*, through *honeſt Sweden*, and *frozen Lapland*, *rude* and *churliſh Finland*, *unprincipled Ruſſia*, and the *wide ſpread regions* of the *wandering Tartar*, if hungry, dry, cold, wet, or ſick, the women have ever been friendly to me, and uniformly ſo; and to add to this virtue, (ſo (ſo worthy the appellation of benevolence) theſe actions have been performed in ſo free, and ſo kind a manner, that if I was dry, I drank the ſweeteſt draught; and if hungry, I eat the coarſe morſel with a double reliſh."

But though the native benevolence, which even among Savages diſtinguiſhes and adorns the female character, might ſometimes ſoften the ſeverity of his ſufferings, yet at others he ſeems to have endured the utmoſt preſſure of diſtreſs.

" I am accuſtomed---(ſaid he, in our laſt converſation---'twas on the morning of his departure for Africa)---I am accuſtomed to hardſhips. I have known both hunger and nakedneſs to the utmoſt extremity of human ſuffering. I have known what it is to have food given me as charity to a madman; and I have at times been obliged to ſhelter myſelf under the miſeries of that character, to avoid a heavier calamity. My diſtreſſes have been greater than I have ever owned, or ever *will* own to any man. Such evils are terrible to bear; but they never yet had power to turn me from my purpoſe. If I live, I will faithfully perform, in its utmoſt extent, my engagement to the Society; and if I periſh in the attempt, my *honour* will ſtill be ſafe, for death cancels all bonds."

CHAPTER III.

Arrival of Mr. LUCAS *at Tripoli.---His Reception by the Baſhaw.---His Journey to Meſurata with the Shereefs Fouwad and Imhammed.---His Mode of obtaining from the latter an Account of his Travels in the Interior Countries of Africa.---His Return to England.*

MR. LUCAS, having taken his paſſage at Marſeilles, on board the St. Jean Baptiſte, a ſmall veſſel belonging to that port, embarked on the 18th of October, 1788; and on the 25th of the ſame month arrived in the harbour of Tripoli.

The date trees, which ſpread themſelves like a foreſt behind the town, and the hills beyond them, which bound the proſpect on the South, are intereſting objects; but the town itſelf is built in too low a ſituation to compoſe a part of the general ſcene: for it is ſcarcely viſible at the diſtance of a mile.

The firſt appearance of Tripoli may diſappoint, by its meanneſs, the expectations of the traveller; but if he reflects on the nature of a deſpotic government, ever incompatible with permanent proſperity, he will not be ſurprized when he finds, on a nearer view, that the city, though the capital of an empire, exhibits through all its extent, the marks of a rapid decay; that its ſcanty limits, though ſcarcely four miles in circumference, are too great for its preſent population; and that its antient caſtle, though

though once the pride, and ſtill the reſidence of the reigning family, is now a mouldering ruin.

The expected ceremonial of announcing to the Baſhaw, which is the title of the Sovereign, and to the Conſul of the State, to whom the veſſel belongs, her arrival in the harbour, having been regularly obſerved, Mr. LUCAS, accompanied by Mr. Tully, the Britiſh Reſident, waited on Hadgee Abdrahaman, the Tripoline Miniſter for Foreign Affairs, who had formerly reſided in England as Ambaſſador from the Baſhaw; and having known Mr. LUCAS there, received him now with the joy of an old acquaintance, and the cordiality of an intimate friend. Encouraged by this kindneſs, Mr. LUCAS explained to him the object of his miſſion, and requeſted that he would preſent and recommend

 him

him to the Baſhaw, and to the Prince, his eldeſt ſon, who is diſtinguiſhed by the title of the Bey. The Miniſter conſented; and the next morning was, accordingly, appointed by the Baſhaw for the firſt of theſe audiences: the morning after was fixed on by the Bey for the latter.

The Baſhaw, a ſhort and robuſt old man, of a fair complexion, a pleaſing countenance, and an affable, joyous diſpoſition, received Mr. LUCAS with great complacency, and accepted, with much ſatisfaction, his preſent of a pair of double-barrelled piſtols, mounted with ſilver; but expreſſed his ſurprize, when leave was aſked to viſit his kingdom of Fezzan: for the journey, he ſaid, had never been attempted by a Chriſtian. Mr. LUCAS replied, that he was led to undertake it by the report which he had heard of various Roman antiquities

in

in different parts of the kingdom, and by the hope of collecting a variety of medicinal plants that are not to be found in Europe. The Bafhaw appeared to be fatisfied, and promifed that, on the firft opportunity of a fafe conveyance, he would give him fuch aids for the journey as his countenance and protection could afford.

On the next morning Mr. LUCAS was prefented to the Bey, the Bafhaw's eldeft fon, a tall and well fhaped, but dark complexioned man, in the middle period of life; and was received by him with the engaging politenefs for which he is eminently diftinguifhed. The prefent that was made to him, except that its value was inferior, was fimilar to that which had been given to his father; and the affurances of the protection and friendfhip, which he offered him in return,

turn, were the ſame in effect with thoſe which the Baſhaw had expreſſed.

Soon after his preſentation at Court, Mr. LUCAS was informed that ſome of the principal Tribes of the tributary Arabs had lately revolted from the Government, and were then in actual rebellion; that all the frontiers of Tripoli, on the ſide of the Deſart, were infeſted by their inroads; that a caravan from the inland country had lately been attacked; and that a Spaniſh Merchant had been plundered within a few miles of the Capital. Mr. LUCAS was alſo informed, that the Baſhaw, who has no regular forces, was preparing to raiſe, on this occaſion, an army of 2,000 men; that as ſoon as the graſs ſhould be high enough to afford the neceſſary forage for the cattle, which it would be in the month of December, they would begin their

their march to the frontier, where they would be joined by the troops of ſuch of the Arabs as continued faithful to the Government.

With this army, the collective numbers of which were expected to amount to five or ſix thouſand men, it was hoped that the Bey, by the uſual enforcements of predatory war, would be able to reduce the rebellious Tribes to their antient obedience, and to the payment of the cuſtomary tribute.

But while, from this expectation, Mr. Lucas waited with impatience for the departure of the army, he was informed that two Shereefs from Fezzan, who were both, as their title announces, deſcended from the Prophet, and one of whom had married the daughter of the King, were arrived in Tripoli. They

came

came there as Merchants, and brought with them, for ſale, a variety of articles, of which ſlaves and ſenna were the chief: and as the reverence in which the deſcendants of Mahomet are held ſecures their perſons from violence, and their property from plunder, they did not think that the reſtoration of peace was requiſite for the ſafety of their return. It was, therefore, with much ſatisfaction that the Miniſter, whoſe intimate acquaintance they were, received from them an aſſurance, that if Mr. LUCAS could bear the fatigue of the journey, they would take him under their protection, and would be anſwerable for his ſafe arrival in Fezzan.

The next morning, in conſequence of this converſation, the Shereefs waited upon Mr. LUCAS. One of them, whoſe rank as ſon-in-law to the King, entitled him

him to the firſt conſideration, was a tall, thin, copper complexioned man, of too ſlender a frame for his height, which was nearly ſix feet, but of an appearance that was expreſſive of dignity: to this appearance the ſedateneſs of his manners, and the fewneſs, but ſolidity of his words were particularly ſuited. His age was ſeemingly about thirty-five years, and his name was Mahommed Benſein Haſſen Fouwad. The other Shereef was a lively old man, ſhort and thin, and dark coloured, almoſt to blackneſs; affable, free, and entertaining in his converſation, and much reſpected by his companion, to whom he was related. His name was Imhammed, and his age about fifty years.

After many compliments, for which their countrymen are famous, they expreſſed to Mr. LUCAS the pleaſure they ſhould

ſhould feel in preſenting him to their King, who had never ſeen a Chriſtian Traveller, and would be highly gratified by ſo new a viſit. They aſſured him of every accommodation which their country could afford, and of every proof which they themſelves could give of the kindeſt good will, and of the ſincereſt friendſhip. The conference was concluded by a preſent from Mr. LUCAS of a pair of piſtols to each, with a ſuitable quantity of powder and ball and flints.

The Baſhaw, being informed by the Miniſter of the propoſal and promiſes of the Shereefs, expreſſed his approbation of the ſcheme, and ſent, from his ſtables, as a preſent to Mr. LUCAS, a handſome mule for the journey. The Bey, too, was no ſooner acquainted with the arrangements, than he gave directions to a Jew Taylor, who had been employed

in

in making, and had juſt finiſhed his own tent, to wait upon Mr. LUCAS, and take his orders for ſuch a tent as would be requiſite for his journey.

But while in this manner Mr. LUCAS was preparing for his departure, and had beſpoken a Turkiſh dreſs for himſelf, and a magnificent robe, as a preſent, to the King of Fezzan, an apprehenſion aroſe in the mind of the Baſhaw, that if Mr. LUCAS ſhould be taken priſoner by the Rebels, he himſelf ſhould be reduced to the diſtreſſing dilemma of either concluding a diſadvantageous peace, or of abandoning the Interpreter of the King of Great Britain to all the inſults, and all the cruelties which thoſe Barbarians might be diſpoſed to inflict.

For this reaſon, the force of which will be much more apparent, if the re-

ſpect

ſpect in which the office of Interpreter in a Mahometan Government is uſually held, be conſidered, he expreſſed his deſire (and in this deſire his eldeſt ſon, the Bey, entirely concurred) that Mr. LUCAS would defer his intended journey till the revolted Arabs ſhould be reduced to obedience, and the peace of the Deſart be reſtored. A few days after this requiſition, the Bey began his march with an army of 300 horſe and 1500 foot.

The Shereefs were no ſooner informed of the obſtacle which had ariſen to the journey of their intended fellow-traveller, than they expreſſed as much chagrin and diſappointment as Mr LUCAS himſelf could feel; for they ſaid, that they had already ſent word to their Sovereign, that they ſhould ſoon have the pleaſure to preſent to him a Chriſtian, who

who had travelled from his native land, (a journey of many moons) with no other view than to gratify his wiſh to viſit him, and to ſee his kingdom of Fezzan:—that his anger would fall heavily on them, to whom he would attribute the diſappointment; and would probably lead him to inflict on them the greateſt indignity that Shereefs can endure, that of having duſt heaped upon their heads.

Impreſſed with theſe apprehenſions, the Shereefs waited upon the Baſhaw, and offered to be reſponſible with their lives for the ſafety of the Chriſtian.

In this unfavourable ſtate of Mr. Lucas's proſpect, an old man of the claſs of Maraboots (a name which is given to perſons of diſtinguiſhed ſanctity) informed the Minifter, with whom he had been long acquainted, that he meant, in a few days, to take his departure for Fezzan; and that as the Rebels, in confequence of the march of the Bafhaw's forces, had removed from that part of the country through which he intended to pafs, he would engage that, under his conduct, Mr. LUCAS fhould travel in fafety.

With this propofal Mr. LUCAS, by the advice of the Minifter, and with the confent of the Bafhaw, had determined to comply, though againft his own opinion; for the countenance and behaviour of the Maraboot had fuggefted fufpicions of his fincerity: but while he was preparing for his departure, which was fixed for the Monday following, the Bafhaw, on farther reflection, concluded that

that the plan which the Shereefs had propofed would, on the whole, be attended with the fmalleft hazard.

The fcheme of the journey being thus finally fettled, the Bafhaw, at the requeft of the Minifter, prefented Mr. LUCAS with a letter of recommendation to the King of Fezzan, of which the following is a tranflation.

TRANSLATION OF THE BASHAW OF TRIPOLI'S RECOMMENDATION OF MR LUCAS TO THE KING OF FEZZAN.

" Praife be unto the Almighty God, and unto our Lord his Prophet Mahommed, whofe protection and mercy we crave, and refign ourfelves to his holy will: to our Son Sydy Hamed Benmohamed, the great and juft ruler over his beloved people; may his days be long and happy. Amen.

"Peace, and the protection and blessing of God, be with you, and preserve you from evil.

"We have to acquaint you, our son, that our friend, the English King, hath sent one of his Interpreters unto us, and desired we would procure him a safe conveyance to Fezzan, where he goes for his own amusement and pleasure: and as we have found a person whom we esteem, and who has promised us to take great care of him, we have consented to let the said Interpreter and his friends * go with him to Fezzan. We have

* *In the Arabic, the word Friend is often employed, as in this passage, to express the same meaning as the English term Servant.*

have to desire that you will shew him and his friends every kindness in your power, and comply with all his wishes; and should he be inclined to go to any other place, you will send proper people to conduct him, and to protect him in every thing; for he is a man of sense, and much esteemed by us; wherefore we recommend him to your care and protection. Peace and the blessing of God be with you: from the Slave of God, Ally Benkaramaly, whose greatness is under the protection of God. Dated in the Moon of Rabeah thénee 1203"—(which corresponds with the month of January 1789.)

To this rout by Mesurata, though not so direct as the antient passage by the way of the Mountains of Guariano, the Merchants who trade to Fezzan have lately given the preference: for in the firſt place, they avoid the oppreſſive contributions, which, even in time of peace, the rapacious tribes of Hooled Benſoliman and Benioleed, who inhabit thoſe hills, have often levied on travellers; and in the next place, they have not only the advantage of ſending their heavy merchandize to Meſurata by ſea, but have alſo an opportunity of hiring there, at a much lower rate than at Tripoli, the camels for which they have occaſion.

On Sunday the firſt of February, 1789, at half an hour after eight in the morning, the Shereefs, accompanied by Mr. Lucas, took their departure from the ſuburbs of Tripoli, where, in a garden which is ſituated at the diſtance of three miles from the town, and which belonged to a Tripoline Merchant, who was travelling with them to Fezzan, they and their attendants had ſlept the evening before.

The caravan was compoſed of the Shereef Fouwad, and of three other Merchants, on horſeback, all of them well armed; of the little old Shereef, who rode upon an aſs; of Mr. LUCAS, who was mounted on the mule which the Baſhaw had given him; of Mr. LUCAS's black ſervant, well armed, upon a camel; of twelve Fezzaners on foot, but armed; of three Negros and their wives, who had been ſlaves at Tripoli, but having obtained their freedom, were now travelling to Fezzan on their return to their native country; and of twenty-one camels, with fifteen drivers, each of whom was armed with a muſket and a piſtol.

That ſo few camels were requiſite in this part of the journey, was owing to the expedient which the Shereefs, with great œconomy, had adopted, of ſending their heavy merchandize by ſea to Meſurata.

At twelve o'clock, the caravan, whoſe courſe was E. S. E. paſſed through the town of Tajarah, a miſerable collection of clay-walled huts, of which ſome were covered with terrace, and the reſt with roofs of thatch: but wretched as the buildings are, the country around them abounds with Date trees, among which a few of the Olive are intermixed.

At five the caravan encamped for the night upon a ſandy eminence. No ſooner were the camels unburthened of their loads, than their drivers turned them looſe to feed on the ſtubble of the valleys,

valleys, and on the brambles of the adjacent hills; but though their freedom is thus given them, they never ſtray to a greater diſtance than that of two or three hundred paces from the camp.

The loads in the mean time are piled in a circle, and, except at the narrow opening which forms the entrance, are ſtowed as cloſe as poſſible to each other. Within this circle the Merchants and drivers and ſervants ſpread their mats and carpets. Here, alſo, they light their fires and dreſs their victuals; and without any other covering than their alhaiques or blankets (for very few are furniſhed with a tent) lie down amidſt the heavy dews and occaſional ſtorms of rain that fall upon the coaſt, and ſleep as ſoundly as in bed: for the wetneſs of their cloaths, which is often the conſequence of this expoſure, is little regarded, ed, and from the ſalubrity of the climate, is attended with little inconvenience.

Mr. LUCAS's tent being ſpread, the two Shereefs, with three of their friends, took up their quarters with him: and on the firſt appearance of ſupper, which was ſerved in a large wooden diſh, and conſiſted of dried meat, and of flour formed into balls, and dreſſed in ſteam, they all ſat down with the familiarity of near relations, and dipping their right hands into the diſh, without either ſpoons or forks or knives, devoured, with a voracious and diſguſting haſte, the whole that was ſet before them.

The concluſion of the meal was followed by the ceremony of waſhing, which conſiſted in each man's dipping his right hand into the ſame water which his

compa-

companions had ufed. Coffee being then brought in, they lighted their pipes, and each of them having drank three or four difhes as he fmoaked, they laid themfelves down in their cloaths, upon the bare fand, and converfed together till they talked themfelves to fleep.

February 2d. The next morning, at day-break, the drivers began to re-load the camels: at eight o'clock the caravan was again in motion; from which time till half an hour after four, they travelled amidft dreary hills of loofe and barren fand, where they faw neither man nor beaft, neither wood nor water.

A fmall valley between the hills, from which, to their great annoyance, the fhifting fand was continually blown down upon them, was the place of their encampment; a place entirely deftitute of water;

water; but from this circumſtance they felt no ſort of inconvenience, as they had brought with them, in goat ſkins, an ample ſtore.

February 3d. At half an hour after ſeven in the morning, they proceeded on their journey, and having emerged from the ſand hills about two in the afternoon, were charmed with the ſight of Olive and of Date trees, of large quantities of white thorn, and of the Spaniſh broom; yet the ſoil is dry and ſtoney, and the few fields of grain which preſent themſelves here and there to the eye, exhibit in their ſcanty and meagre appearance, the marks of an ungracious and ſullen vegetation.

On the right or S.E. of their road, at the diſtance of about twenty miles, the mountains of Guariano and Miſſelata riſe upon

upon the view.—A ſight that recalls to the mind of the experienced Traveller, and leads him to relate to the ſtranger, the beauty of the vales, the richneſs of the lands, abounding in corn and oil, and the fierce inhoſpitable diſpoſition of the inhabitants, that compels the caravan to turn from their dominions, its direct and antient road, and take its courſe among the deſolate hills, and dreary waſtes of the ſandy and barren coaſt.*

A requeſt

* Ben Ali, a native of Morocco, who was lately in England, and of whom an account is given in the Introduction to Chapter IV. relates, that in proceeding from Fezzan to Gharien, on his way to Tripoli, he was met by ſeveral parties of Arabs, who were robbers by profeſſion, and who rendered the rout ſo dangerous, that every individual in the caravan was obliged to carry a gun, a brace of piſtols, and a yatagan or ſabre. He deſcribes the country

as

A request from the Shereef Fouwad was now made to Mr. LUCAS for his consent to encamp that evening in the neighbourhood of an old Arab, his particular friend, with whom he had business to transact, but whose residence was two hours march to the Southward of their road. They accordingly turned to the South, and about five o'clock, after a tedious and difficult passage among rocky hills, they approached the tents of the Arab. The old gentleman, accompanied by his two sons and few attendants, came forward to meet them; and after expressing great satisfaction at the

as partially cultivated; and remarks, that it is furnished with few springs, and is wholly destitute of rivers.

By his account, the distance from Fezzan to Gharien is that of a journey of sixteen days.

He represents the rout from Gharien to Tripoli as a sandy desart, and its length as that of a seven days journey.

the ſight of his friend, the Shereef, he ordered a tent to be cleared for their reception, and in the mean time conducted them to a mat and carpet, which his ſervants had ſpread for them under a hedge; for, notwithſtanding the ſeaſon of the year, the heat was already troubleſome. They had not been ſeated long when their hoſt invited them to their tent, in which a number of mats and carpets were neatly laid.—A ſheep was killed, and ſent to be dreſſed for their ſuppers; bowls of butter-milk were brought for their preſent refreſhment, and barley in abundance was given to ſuch of their cattle as were accuſtomed to that kind of food; while the camels, as uſual, were ſent to feed among the hills.

At eight o'clock the ſupper was brought to the tent, and was placed before them

in

in two large wooden diſhes. Of theſe the firſt contained the mutton, which was boiled, and cut into ſmall pieces: the other was filled with a boiled paſte of dried barley meal, made up in the form of an Engliſh pudding, and ſurrounded with a great quantity of oil. This diſh, which was intended as an accompaniment to the mutton, and which is in much eſtimation at Tripoli, is called bazeen.

While Mr. LUCAS taſted of the laſt, and eat with pleaſure of the firſt of theſe diſhes, and the Fezzaners, with their uſual diſpatch, were devouring the contents of both, the old man and his ſons ſtood by to ſupply them with water and butter-milk; for the rules of the Arab hoſpitality require, that during their meals the maſter of the houſe ſhould wait upon his gueſts.

Feb.

Feb. 4th. The next morning, at ſeven o'clock, the entertainment was repeated, with the ſame marks of a kind and liberal welcome; for the old man is rich in corn and cattle, and having obtained the character of a Muſſelman Saint, or Maraboot, is, on that account, exempted from the payment of taxes.

After a march of three hours, during which the route was perplexed, and the eye fatigued by a continued ſucceſſion of rocky hills, the caravan arrived at the entrance of an extenſive and beautiful plain, that every where exhibited a luxuriant growth of Olive trees, intermixed with Dates.

The next two hours brought them to the ſea coaſt, and to all that now exiſts of the town of Lebida, where, in the ruins of a temple, and in the much more perfect remains of several triumphal arches, the Traveller contemplates the magnificence of an antient Roman colony; and discovers, in the beauty and fertile appearance of the adjoining plain, the reasons which led them to chuse, for a sea-port town, a situation that furnishes no natural harbour.

Eastward of the ruins, for about five and twenty miles, the soil, though entirely unaided by the poor Arabs who inhabit it, exhibits the same luxuriant vegetation; and the scene is rendered still more interesting by the remains of a stupendous aqueduct, which formerly conveyed to Lebida the water of a distant hill.

At half an hour after five, and in the neighbourhood of a miserable village, the caravan encamped for the night.

Feb.

Feb. 5th. The next day's journey, which was attended with nothing remarkable, and during which they followed the line of the coaſt, brought them to Zuleteen, an inconſiderable town, where they found that a boat, to which a part of their baggage was intruſted, had been compelled by a ſtorm to depoſit her cargo. From this circumſtance, and the neceſſity which followed it, of hiring ſix additional camels for their goods, the departure of the caravan on the next day (February 6th) was retarded till two o'clock in the afternoon. At the end of the firſt hour's march, they were informed by ſome friendly Arabs, who were moving their tents and cattle, for the ſake of protection, to the ſuburbs of the town, that on the preceding afternoon a party of the rebel tribe of Hooled Benſoliman, from the neighbouring hills, had attacked a ſmall caravan belonging to Meſurata, and after killing four of the people, had carried off the camels and baggage:—and they were alſo informed, that on that very morning two men, who were going from Meſurata to the market, which is held at ſome diſtance from the town, were robbed and killed by the ſame party. At this news a Council was ſummoned to determine on the prudence of attempting to proceed; for the Shereefs began to diſtruſt the ſufficiency of that title to an exemption from the violence of war, on which, when the danger was diſtant, they had ſo confidently relied. The opinion of Mr. Lucas being aſked, he obſerved, that as the party which committed the depredations were deſcribed as not more than forty or fifty in number, and were conſequently much too weak to reſiſt the detachment that, they muſt be ſure, would

would be ſent from Meſurata to revenge the violence of which they had been guilty, he had not the ſmalleſt doubt of their being already returned to the refuge of their mountains; but that at any rate, their own numbers, conſidering how well they were armed, were amply ſufficient to defend them from the attacks of ſuch petty marauders. Pleaſed with an opinion which gave them the proſpect of but little danger, they freſh primed their muſkets and piſtols, and ſinging as they went, drove merrily on.

At ſix o'clock they encamped upon a hill directly oppoſite to the enemy's mountains, that were now within twelve or fifteen miles; and having lighted, by Mr. LUCAS's advice, about ſeventy fires, for which the dry bruſh-wood that was near them furniſhed the means, they had ſoon the ſatisfaction of obſerving, that the fires of the enemy, who probably miſtook them for the troops of Meſurata, were all extinguiſhed.

Feb. 7th. The next morning at day-break, in the midſt of a ſtorm from the S. W. of violent rain and wind, they left the hill; and after a tempeſtuous march of four hours, they diſcerned through the heavy atmoſphere, which now began to clear, a party of fifty or ſixty Arabs upon a riſing ground, at a diſtance to the left.—That more were concealed behind the hill, they had not the ſmalleſt doubt; but as eſcape was impoſſible, and conſultation uſeleſs, they reſolved unanimouſly to *make* rather than *receive* the attack. The Shereef Fouwad took the command, and having given the charge of the camels to the three Negros and their wives, with orders to drive them ſlowly, and keep them cloſe together,

ther, led on the reſt of the party. The horſe, with the Shereef at their head, formed the van, while thoſe on foot were mixed together in a croud, dancing and ſhouting, and twirling their muſkets over their heads, and running round each other like madmen, till they came within ſhot of their antagoniſts, when they ſuddenly diſperſed, and each man ſquatted down behind a buſh, to ſhelter himſelf and take a ſurer aim. The horſe were now cloſe upon the enemy, and were levelling their pieces at the foremoſt, when one of the latter laid down his muſket, and called to them not to fire, for they were friends.

A moment's pauſe was followed by a mutual recollection, and they exhibited, on both ſides, the moſt extravagant marks of joy. They ran round each other like a flock of frighted ſheep, and danced, and ſhouted, and twirled their guns over their heads, till they were tired, when they ſat down and began a reciprocal congratulation on their ſafety. The ſtrangers ſaid that they were herdſmen belonging to Meſurata; that for want of paſturage near the town, they had brought their flocks to feed upon theſe hills; that they were 200 armed men, and that they did not fear the enemy.

After this information, and the exchange of civilities, the caravan continued its journey, and at ſix in the evening arrived at Meſurata.

The Governor, whoſe politeneſs and natural good ſenſe had been improved by a long reſidence in Italy, received Mr. LUCAS with marks of the greateſt attention; but expreſſed his fear that, while

while the war continued, the Shereefs would not be able to obtain from the Rebel Arabs, who alone could furnish them, the 120 camels which were requisite for the conveyance of their goods: and that, as the prospect of peace was at present remote, and the sultry season would soon commence, he saw but little chance of their reaching Fezzan before the following Winter.

Feb. 9th. Information was now received at Mesurata, that the Bey's army, which consisted of 1,500 horse, and 6,000 foot, was encamped within five hours march of the Rebels, whose force was composed of 600 horse and 10,000 foot, and was commanded by a powerful Chief of the name of Séife Bannazar.

It was also said, that the Tribes of the friendly Arabs, who formed the principal

cipal part of the Tripoline army, were too closely connected, by intermarriages and the force of antient alliance, with many of the Rebel Clans, to bring with them to the battle that sort of zeal which Government could safely trust.

Feb. 10th. Such was the situation of affairs when the Shereef Fouwad requested from the Governor, to whom he was strongly recommended by the Minister, a public and formal declaration, that if the hostile Arabs would send to Mesurata 120 camels, with their drivers, for the conveyance of the merchandize of the Shereefs to Fezzan, both they and their catttle should be perfectly secure. The Governor replied, that by his own authority alone he could not, with either prudence or effect, announce to the Rebel Arabs such a stipulation; but that he would summon a Council of the

the Chiefs of the town, and would propose the business to them; though he himself was persuaded, that should they consent to the Shereef's requisition, as he hoped they would, the Rebel Arabs were much too cautious to rely on the good faith of such an engagement.

Feb. 11th. The next morning, a Council of six of the principal inhabitants, with the Governor as President, assembled in Mr. LUCAS's tent, (for the Governor's own house was near the sea, at the distance of six miles from Mesurata) and unanimously agreed that a letter should be written by the Governor, and signed by himself and by all the Members of the Council, to assure the hostile Arabs, that such of their camels and of their people, as they might send at the request of the Shereef, should neither be detained or molested within the

the jurifdiction of Mefurata. This letter, accompanied by one from the Shereef, in which he defired to be furnifhed with 120 camels for the carriage of his goods, was accordingly fent on that very day, by an exprefs, to a rebel province, in which he had many friends, and which is called Guady.

Feb. 14th. In three days from the time of his departure, the exprefs returned, and brought with him a reply, in which the Arabs obferved that, as the country was in arms they could not with prudence truft their camels from under their own protection, much lefs could they fpare their people.

Notwithftanding this anfwer, the Shereef Fouwad conceived that the refufal of the Arabs was folely dictated by a diftruft of the fort of fecurity which was offered

offered by the Governor and Council of Mesurata; for independently of the doubts which the Arabs might entertain of their good faith, it was evident, that without the sanction of the Bey, who commanded the army, their engagement, at the utmost, could not extend beyond their own jurisdiction. But if the Bey himself would guaranty the safety of the camels and their drivers, by granting them a pass, the Shereef concluded that the real objection of the Arabs would be entirely removed. With this view, on the 27th of February, the Shereef and two of his countrymen set out for the camp, which they reached on the second day, as it was not far from Mesurata; but their trouble was fruitless, for the Bey could not be prevailed on to assent to their proposal.

All hopes of obtaining, before the con-

conclusion of the war, a sufficient conveyance for the goods being thus at an end, the Council resolved that, until peace should be established, the Shereefs and the other Merchants of the caravan, should be at liberty to warehouse their packages in the public store-rooms of the Governor.

Deprived, in this manner, of all prospect of arriving this year at Fezzan, and doubtful if the state of the country would encourage, or his own situation permit the attempt in the Winter, Mr. Lucas resolved to avail himself to the utmost of such means of information as the knowledge of his fellow-travellers enabled them to afford.

He had already discovered that the little old Shereef Imhammed had been often employed by the King of Fezzan

as

as his Factor in the Slave Trade; and in that capacity had travelled to Bornou and different parts of Nigritia; and he now determined to cultivate his friendſhip with double ſolicitude, and by occaſional preſents and frequent converſation, to draw from him an account of the countries which he had ſeen. With this view he, one evening, took from his pocket his map of Africa, and after ſatisfying the Shereef's curioſity as to its nature and uſe, told him that he once intended it as a preſent to the King of Fezzan; but, that having diſcovered in it ſeveral miſtakes, he now propoſed to draw another that ſhould be more correct. The Shereef replied, that the King would be highly pleaſed with ſuch a preſent. Mr. LUCAS ſaid, that if he would aſſiſt him with an account of the diſtances from place to place, in ſuch parts of the country as he had viſited, and with their

names

names in Arabic, and would alſo ſatisfy him as to ſuch queſtions as he ſhould aſk, he would prepare *two* corrected copies of the map, and would give one of them to the King and the other to himſelf. The Shereef was delighted with the propoſal; and they immediately retired to a ſand hill at ſome diſtance from the tent, that their converſation might be unreſerved and uninterrupted. Many ſucceſſive days were employed in the ſame manner; and as Mr. LUCAS wrote down, at the time, the information which he obtained, he was ſoon poſſeſſed of ſuch an account of Fezzan, Bornou, and Nigritia, eſpecially of the two former, as much diminiſhed the chagrin of his own diſappointment.

One afternoon, as they ſat together on the cuſtomary hill, they were ſuddenly diſturbed by the loud ſcreams and

diſmal

diſmal howlings of all the women of Meſurata—a mode of alarming and collecting the men, which is always practiſed among the Arabs, on the approach of thieves, or of an invading enemy.

In a few minutes the townſmen were under arms, and together with the Shereef Fouwad, the other Fezzaners, and Mr. LUCAS's Black, went haſtily on to the place where the Rebels were ſaid to have appeared:—there they found that the women had been deceived. It ſeems an aſs had ſtrayed into a field of barley; and as the owner of the corn, who was armed, and happened to paſs by at the time, went into the field to drive out the animal, the women miſtook him for one of the Rebels, and conceiving that many more were concealed, (for they often come down from the mountains to ſteal the cattle) had given the uſual alarm.

In a few minutes, Mr. LUCAS and the old Shereef, who had both continued on the hill, obſerved the Fezzaners coming ſulkily back, and curſing the women for ſo fooliſh a diſturbance, whilſt the townſmen, on the contrary, fired their pieces, and rejoiced in their diſappointment as much as if they had conquered an army.

A few days afterwards, a ſecond alarm was given, and with much more reaſon than the firſt; for a party of the Rebel Arabs, ſome on horſeback, and others on foot, had ſuddenly appeared within two miles of the town, and after killing two herdſmen, and ſeizing three Black ſlaves, their aſſiſtants, had carried off ſixty goats, fourteen cows, and three camels.

The attack was made at a time when

moſt of the townſmen were at the market, which is held at the diſtance of three miles from Meſurata; and to add to their indignation, it was made in a place which hitherto had been deemed inviolable; for the land on which the cattle were feeding was conſidered as under the immediate protection of a departed Saint, whoſe remains were buried there, and whoſe ſanctuary, it was thought, no Muſſelman, however accuſtomed to robbery and blood, could venture to profane.

March 13th. Letters by expreſs from the camp were now received by the Governor, which announced, that in conſequence of the Bey's having entered the country of the Rebels, and turned his cattle to paſture in their corn, an engagement, which ſoon became general, had enſued; that after a loſs of 150 men, the Rebels had retired to the mountains; and that the Bey, at the expence in killed and wounded, of not more than twenty-ſix horſe and ſeventy or eighty foot, had obtained poſſeſſion of ten or twelve thouſand ſheep, and of three hundred camels.

Mr. LUCAS congratulated the Governor on the victory, who thanked him; but "I fear," ſaid he, ſhaking his head, "that the news requires confirmation. There *was* a time, indeed, when the people of Tripoli knew how to conquer, and the Arabs to tremble at the ſight of an encampment."

March 15th. On the next day but one, accounts were brought by different perſons who arrived from the camp, that there had indeed been a ſkirmiſh, in their relation of which they varied much from each other; but they all agreed that the

Bey

Bey had loſt the greater number of men, and that the only cattle which he had obtained, were a few camels and ſome ſheep that the ſtraggling parties from the camp had ſeized.

Wearied with fruitleſs expectations of a peace, diſappointed in their expedients, and warned by the increaſing heat, that the ſeaſon for a journey to Fezzan was already paſt, the Shereefs now reſolved to proceed to the intended places of their Summer reſidence.

The Shereef Fouwad retired to Wadan, his native town; and the Shereef Imhammed, with tears in his eyes, and an earneſt prayer that he might ſee his friend Mr. LUCAS again in November, retired to the mountains, where he had many acquaintance, and could live at a ſmall expence.

March 20th. A few days afterwards, Mr. LUCAS took leave of the Governor, to whofe civilities he had been much indebted, and having accompanied a fmall caravan as far as Lebida, embarked in a coafting veffel at the neighbouring village of Legatah, and went by fea to Tripoli, where the Bafhaw, upon whom he waited, and to whom with many acknowledgments he returned the mule, not only received him with great kindnefs, but expreffed his hope that better fortune would attend him another year.

April 6th. From Tripoli he failed for Malta, and after a tedious quarantine, which the fufpicion of the plague at Mefurata had much prolonged, he took his departure for Marfeilles, and on the 26th of July arrived in England.

INTRO-

INTRODUCTION

TO

CHAPTER IV.

AN account has already been given of the opportunity which the length of his reſidence in Meſurata afforded to Mr. LUCAS, of obtaining from the Shereef Imhammed a deſcription of the Kingdom of Fezzan, and of ſuch of the countries beyond it to the South as the Shereef himſelf had viſited.

But though this intelligent ſtranger had no diſcoverable motive for deception, yet as the ſolitary evidence of any individual excites but a dubious belief, Mr. LUCAS was anxious to learn from the Governor of Meſurata, who had formerly travelled to Fezzan, his idea of the truth of the Narrative. With this view he aſked and received the Governor's permiſſion to read to him the memorandums that the repeated converſations of Imhammed had enabled him to make.

"The Shereef's Account of Fezzan," ſaid the Governor," my own knowledge confirms; and many of the particulars which he relates of Bornou and Caſhna I have heard from the report of others. His countrymen ſay that he is better acquainted with both than any other individual among them; and ſuch is the opinion which the King himſelf entertains of his probity, knowledge, and talents, that to his management is always entruſted whatever buſineſs in either of

thoſe

thoſe empires his Sovereign has to tranſact."

But while Mr. LUCAS, with a prudent and laudable caution, was thus endeavouring to aſcertain the truth of the Shereef's account, another, and perhaps more deciſive teſt of its value was fortunately obtained in England.—For, before the return of Mr. LUCAS, or the arrival of his papers, the Committee of the Aſſociation, aſſiſted by Mr. DODSWORTH, (whoſe reſidence of fourteen years in Barbary had given him a competent knowledge of Arabic) had procured from Ben Alli, a native of Morocco, at that time in London, an account of all thoſe countries to the South of the Deſart of Zahara, which, in the courſe of his extenſive Travels as a Merchant, he had formerly viſited: and though his remarks appear to be thoſe of a ſuperficial

cial Obſerver, who poſſeſſes activity of ſpirit rather than energy of mind, and whoſe remembrance of what he ſaw is impaired by the lapſe of near twenty years; yet, (as will be ſeen in the following pages) the general conformity of his deſcription of Bornou to that which the Shereef has given, has an obvious tendency to ſtrengthen the credit of the latter.

This ſhort account of the nature of the only external evidence that has yet been obtained in ſupport of the following Narrative, ſeemed to be due from the Committee; but in what degree that evidence is impreſſive of belief, or what internal marks of authenticity the Work itſelf may afford, the judgment of others muſt decide; for on theſe points, it is evident that each individual muſt determine for himſelf. In forming his opinion, however,

ever, it is requiſite he ſhould know, that while the moſt anxious attention has been given to the faithful preſervation of the ſenſe of the Original, an entire change has been made in its language and arrangement; a change which, the obvious advantage of methodizing converſations, as deſultory as they were numerous, of ſeparating the blended accounts of unconnected objects, and of uniting a variety of broken and detached deſcriptions of the ſame thing, has unavoidably occaſioned.

CHAPTER

CHAPTER IV.

Rout from Mesurata.---Enumeration of the principal Towns of Fezzan---Account of its Climate and principal productions---Description of the Manners, Religion, and Government of its People, their Revenue, Administration of Justice, and Military Force.

FEZZAN, whose small and circular domain is placed in the vast Wilderness, as an island in the midst of the ocean, is situated to the South of Mesurata. A journey of eight days, through districts but little inhabited or improved, though naturally not unfertile, conducts the Traveller to the town of Wadan, where every

every requiſite for the refreſhment of the caravan is found.—From thence, in five hours, he arrives at the forlorn village of Houn, on the edge of the Deſart of Soudah, on whoſe black and obdurate ſoil, the baſis of which is a ſoft ſtone, no vegetable but the Talk is ſeen to grow. To this tree, which is of the ſize of the ſmall Olive, and bears a ſprig of yellow flowers, the huſbandman of Fezzan is indebted for the hard and lemon-coloured wood of which he forms the handles of his tools, and the frames of his larger inſtruments. Having croſſed the Deſart, which furniſhes no water, and for the paſſage of which four days are requiſite, the Traveller accepts the refreſhments of a miſerable village that affords him nothing but dates of the worſt quality, ſome brackiſh water, and a ſmall ſupply of Indian corn, of the ſpecies called Gaſſôb. From Zêghen, by which name the

village

village is diſtinguiſhed, a ſingle day conducts him to the town of Sebbah, where the large remains of an antient caſtle, built upon a hill, and other venerable ruins, that in point of extent are compared to thoſe of Lebida, impreſs on his mind the melancholy idea of departed greatneſs; while, on the other hand, the humble dwellings of the modern inhabitants, and the rich vegetation of their neighbouring fields, preſent to his eye an ample ſtore of all that is requiſite for the ſuſtenance of man.—Dates, barley, Indian corn, pompions, cucumbers, fig trees, pomegranates, and apricots, and for meaner purpoſes, the white thorn and Spaniſh broom are deſcribed as but a part of the numerous vegetables that reward the induſtry of the people. The animals in which they moſt abound are ſaid to be the common fowl,

and

and the brown long-haired and broad-tailed ſheep.

From Sebbah a journey of two days tranſports the Traveller to Goddoua, a ſmall town of ſimilar produce; and from thence, in two days more, he arrives at Mourzouk, the capital of the kingdom of Fezzan.

This city* is ſurrounded by a high wall, which not only furniſhes the means of

* The capital of Fezzan is ſituated on the banks of a ſmall river, and is alſo ſupplied with water from a multitude of ſprings and wells.

Being formerly built with ſtone, it ſtill retains the appellation of a Chriſtian Town; and the medley which it preſents to the eye, of the vaſt ruins of antient buildings, and of the humble cottages of earth and ſand that form the dwellings of its preſent Arab inhabitants, is ſingularly groteſque and ſtrange.

BEN ALLI.

of defence, but affords to the Government an opportunity of collecting, at its three gates, a tax on all goods (though provisions are exempted) that are brought for the supply of its people. Its distance from Mesurata, which borders on the coast, and with respect to which its situation is nearly South, is about* 390 miles.

Eastward of Mourzouk, and situated in a district of remarkable fertility, is

K the

* *In this estimate of distance, the rate of travelling is supposed to be twenty-two miles per day:—a conclusion that arises from the time that was employed by* Mr. Lucas *in travelling from Tripoli to Mesurata; for in that journey of* 150 *miles, seven days were consumed; and though the caravan was detained for a few hours on the sea coast, and was employed during four more in passing to and from the tents of the Arab, yet these losses were probably compensated by the extraordinary dispatch with which, in consequence of their fears, the greatest part of their journey was performed.*

the town of Zuéela, in which the remnants of antient buildings, the number and ſize of the ciſterns, and the conſtruction of the vaulted caves, intended perhaps as repoſitories for corn, exhibit ſuch veſtiges of antient ſplendour, as will probably attract, and may highly reward the attention of the future Traveller.

To the South of Zuéela, and nearly at the ſame diſtance from the capital, is the town of Jermah, which, like Zuéela, is diſtinguiſhed by the numerous herds, eſpecially of ſheep and goats, that are ſeen around it; by the various and abundant produce of its adjacent fields; and by numerous and majeſtic ruins, that exhibit to the ignorant inhabitants of its clay-built cottages inſcriptions of which they know not the meaning, and veſtiges

veſtiges of greatneſs to which they are perfectly indifferent.

Teſſouwa, a conſiderable town, is alſo ſituated to the Eaſtward of the capital; but ſeems to have no claim to particular attention. Near this town, a river which the Shereef deſcribes as overwhelmed in the moving ſands, but which he remembers a deep and rapid ſtream, had formerly its courſe.

More remote from Mourzouk, being diſtant from it in a N. E. direction, about 120 miles, is the large town of Temmiſſa. Here the caravan of Pilgrims from Bornou and Nigritia, which takes its departure from Mourzouk, and travels by the way of Cairo to Mecca, uſually provides the ſtores of corn and dates, and dried meat, that are requiſite for its dreary paſſage,

S. E. from the capital, and diſtant from it about ſixty miles, is the ſmall town of Kattrón, which ſeems to be remarkable for nothing but the quantity which it breeds of the common fowl, and for the abundant crops of Indian corn which the neighbouring lands afford.

Very differently diſtinguiſhed is the town, or rather the province, of Mendrah, for though much of its land is a continued level of hard and barren ſoil, the quantity of *Trona*, a ſpecies of foſſil alkali that floats on the ſurface, or ſettles on the banks of its numerous ſmoaking lakes, has given it a higher importance than that of the moſt fertile diſtricts.

Of this valuable produce great quantities are annually brought by the Merchants of Fezzan to Tripoli, from whence it is ſhipped for Turkey and

Tunis,

Tunis, and the dominions of the Emperor of Morocco. The people of the latter employ it as an ingredient in the red dye of the leather, for which they are famous, and in that of the woollen caps that are worn by the Arabs and the Moors as the baſis of their turbans.

The ſituation of Mendrah is nearly South from the capital, and is diſtant from it about ſixty miles.

To the account that has been given of the principal towns of Fezzan, that of Tegérhy alone remains to be added. It is but a ſmall town, is ſituated S. W. of the capital, about eighty miles, and collects from its lands but little other produce than dates and Indian corn. The territory of Fezzan, to the Weſtward of the capital, appears to extend but a little way; for on that ſide, the ſullen barrenneſs of the Deſart, more effectually than the ſtrongeſt human power, preſcribes a limit to the purſuits of Avarice and to the efforts of Ambition.

Of the ſmaller towns of Fezzan, and of its ſcattered villages, the number of which, including that of the towns, is ſaid to be little leſs than one hundred, the Shereef has given no particular deſcription.

The towns themſelves appear to be chiefly inhabited by huſbandmen and ſhepherds; for, though they alſo contain the Merchants, the Artificers, the Miniſters of Religion, and the Officers of the Executive Government; yet, the buſineſs of agriculture and paſturage ſeems to be the principal occupation of the natives of Fezzan.

In

In every town a market for butcher's meat, and corn, and fruit, and garden vegetables, is regularly held. Mutton and goats flesh are sold by the quarter, without being weighed: the usual price of a quarter of a goat or sheep is from thirty-two to forty grains of gold dust, or from four to five shillings of English money. The flesh of the camel, which is much more highly valued, is commonly sold at a dearer rate, and is divided into smaller lots.

The houses, like those of the little villages in the neighbourhood of Tripoli, are built of clay, and are covered with a flat roof, that is composed of the boughs and branches of trees, on which a quantity of earth is laid. Inartificial and defective as this covering appears, it is suited to the climate: for as rain is never known in Fezzan, the principal requifites of a roof are fhelter from the dews, and protection from the fun.

The heats of the Summer, which begins in April and continues till November, are fo intenfe, that, from nine in the morning till fun-fet, the ftreets are frequented only by the labouring people, and even in the houfes refpiration would be difficult, if the expedient of wetting the apartments did not furnifh its falutary aid. Of this torrid clime the fiercenefs is chiefly felt from the month of May to the latter end of Auguft; during which period, the courfe of the wind is ufually from the E. the S. E. the S. or the S. W. and though from the two latter points it blows with violence, the heat is often fuch as to threaten inftant fuffocation: but if it happens to change, as, for a few days, it fometimes does, to the W,

W. or N. W. a reviving freſhneſs immediately ſucceeds.

The dreſs of the inhabitants of Fezzan is ſimilar to that of the Moors of Barbary. The immediate covering of the body conſiſts of a pair of large trowſers, of linen or cloth, which deſcends to the ſmall of the leg, and of a ſhirt, which is wide in the ſleeves, but looſe at the breaſt, and the ſkirts of which hang over and conceal the upper part of the trowſers. Next to the ſhirt is worn a kind of waiſtcoat, which in ſhape reſembles the ſhirt, except that it has no ſleeves, and that it reaches no lower than the waiſt; and to the waiſtcoat is ſuperadded a jacket, with tight ſleeves which extend to the wriſt, but which are left unbuttoned and open from the wriſt almoſt to the elbow. Thus far their dreſs may be ſaid to be ſimilar to that of a

Britiſh

Britiſh ſeaman, its colour excepted, and except too, that the ſhirt is not open at the breaſt, that the waiſtcoat is not faſtened with buttons, but is put on like the ſhirt, and that the bottom of the ſhirt hangs down on the outſide of the trowſers.

Over the jacket is worn a looſe robe, which reaches below the knee, and the ſleeves of which, though wider than thoſe of the jacket, are made in the ſame form, and, like them, are left open at the wriſt.

A girdle of crimſon ſilk binds the robe to the waiſt; and a long cloth (called a barakan or alhaique) of the ſhape of a Highlander's plaid, and worn in the ſame way, is thrown over the whole. The legs, as far as the calves, to which the trowſers

trowſers deſcend, are covered with a kind of ſhort ſtockings, which are made of leather, and are laced like the half boot of an Engliſhman. The feet are accommodated with ſlippers; and the head is protected by a red woollen cap, which is incircled by the folds of a ſilk or muſlin turban.

Ample as this cloathing may appear, the farther proviſion of a long cloak with a large hood is often conſidered as requiſite. It is called a burnooſe, and in fine weather is uſually carried on the ſhoulder.

Such, when complete, is the dreſs of the inhabitants of Fezzan. But in the Summer months the common people have no other covering than the drawers, which decency requires; and the caps, which protect their heads from the immediate

action

action of the ſun; for in other reſpects their bodies are completely naked.

Nature and cuſtom have formed their conſtitutions to ſuch high degrees of heat, that any approach to the common temperament of Europe entirely deſtroys their comfort; for Mr. LUCAS often obſerved, in his journey to Meſurata, that when the ſcorching heat of the noon-day beams had compelled him to ſeek the ſhade, his fellow-travellers, eſpecially if the wind was in the North, laid themſelves down, upon the ſand in the open ſun, in order to receive a double portion of its warmth; and when, as their cuſtom was, they enquired after his health, they, almoſt always, concluded with the expreſſion, "*Heack m'andick berd,*" we hope you are not cold.

The diſeaſes that are moſt frequent in

Fezzan

Fezzan are thoſe of the inflammatory, and thoſe of the putrid kind.

The ſmall-pox is common among the inhabitants; violent head-achs attack them in the Summer; and they are often afflicted with rheumatic pains.

Their old women are their principal phyſicians. For pains in the head they preſcribe cupping and bleeding; for pains in the limbs they ſend their patients to bathe in the hot lakes, which produce the trona; and for obſtinate achs and ſtrains, and long continued ſtiffneſs in the muſcles, they have recourſe, like the horſe-doctors of Europe and the phyſicians of Barbary, to the application of a burning iron.

The uſe of the ſtrongeſt oils, and of the

the moſt powerful herbs, is alſo frequent among them.

To the nature of their climate the greateſt part of their diſeaſes is probably owing; and to this cauſe they are certainly indebted for the extraordinary multitude of noxious and of loathſome animals that infeſt their country. Adders, ſnakes, ſcorpions, and toads, are the conſtant inhabitants of their fields, their gardens, and their houſes. The air is crowded with moſquitos; and perſons of every rank are over-run with all the different kinds of vermin that attack the beggars of Europe; and though in the Summer the fleas intirely diſappear, the inhabitants are ſcarcely ſenſible of relief.

In their perſons, the natives of Fezzan

zan incline to the Negro much more than to the Arab caſt. Thoſe who travelled with Mr. LUCAS from Tripoli to Meſurata, and who were fourteen in number, had ſhort curly black hair, thick lips, flat broad noſes, and a dark* ſkin, which, either from their habitual naſtineſs, the vermin with which they are covered, or the natural rankneſs of their perſpiration, emits the moſt nauſeous and fetid effluvia. They are tall, but not ſtrong; well ſhaped, yet indolent, inactive, and weak; and though the Shereef Fouwad is deſcribed as majeſtic in his appearance, yet his countrymen, in general, are conſidered at Tripoli as a people of remarkable uglineſs.

In

* The people are of a deep ſwarthy complexion.

BEN ALLI.

In their common intercourſe with each other all diſtinctions of rank appear to be forgotten; for the Shereef and the loweſt plebean, the rich and the poor, the maſter and the man, converſe familiarly, and eat and drink together. Generous and eminently hoſpitable, the Fezzaner, let his fare be ſcanty or abundant, is ever deſirous that others ſhould partake of his meal; and if twenty people ſhould unexpectedly viſit his dwelling, they muſt all participate as far as it will go.

When they ſettle their money tranſactions, they ſquat down upon the ground, and having levelled a ſpot with their hands, make dots as they reckon; and if they find themſelves wrong, they ſmooth the ſpot again, and repeat the calculation. All this time the by-ſtanders, though they have nothing to do with the buſi-

neſs,

nefs, are as eager to put in their word, and to correct miftakes, as if the affair were their own. Even in common converfation, if they fit without doors, they level the fand in order to go on with their difcourfe, and at every fentence mark it with their fingers.

An extenfive plain, encompaffed by mountains, the irregular circle of which is interrupted on the Weft, where it feems to communicate with the Defart, compofes the Kingdom of Fezzan. To the influence of the neighbouring heights it may poffibly be owing, that in Fezzan, as in the Upper Egypt, the fituation of which is extremely fimilar, no rain is ever known to fall.

A light fand conftitutes the general foil; and fand hills of various forms are feen in particular diftricts; but though character of the ſurface and the dryneſs of the Heavens may ſeem to announce an eternal ſterility, yet the ſprings are ſo abundant, and ſo ample a ſtore of ſubterraneous water is ſupplied by the adjacent heights, that few regions in the North of Africa exhibit a richer vegetation. From wells of eight or ten feet deep, with ſeveral of which every garden and every field are furniſhed, the huſbandman waters, at ſun-riſe, the natural or artificial productions of his land. Of theſe the principal are,

The Talk, a tree that in ſize reſembles the ſmall Olive. It flowers in yellow ſprigs, and ſupplies the hard and lemon-coloured wood, from which the handles and frames of the Fezzanner's inſtruments of huſbandry are made:

The White Thorn:

A kind

A kind of bruſhwood that reſembles the Spaniſh broom:

The Date tree, which is common:

The Olive and the Lime, which are deſcribed as ſcarce; the Apricot, the Pomegranate, and the Fig:

Indian corn and barley, the two favourite objects of the Fezzanner's cultivation:

Wheat, of which but little is raiſed:

Pompions or calabaſh, carrots, cucumbers, onions, and garlick.

Of the *tame* animals that are raiſed in Fezzan, the Shereef enumerates,

The Sheep, which is deſcribed as of a light brown colour; as having a broad tail, and as cloathed with a ſpecies of hair rather than of wool:

The Cow, which does not ſeem to be common, except in a few diſtricts in which the paſture is excellent:

The Goat, and the Camel:

A ſpecies of the domeſtic fowl of Europe.

The *wild* animals of the country are,

The Oſtrich:

Antelopes of various kinds, one of which is called the Huadee, and is celebrated for the ſingular addreſs with which, when chaſed by the hunter amidſt its

its craggy heights, it plunges from the precipice, and lighting on its hams, without danger of purſuit, continues till evening in the vale below:

A ſpecies of deer of a ſmaller ſize than the common park deer of England. Its head, neck, and back, are of a browniſh red; and a pale ſtreak of the ſame colour, running on a white ground, is continued on each ſide from the haunch to the hoof: the reſt of the body is of a clear and delicate white. Such, if the Fezzanners are to be credited, is the cleanlineſs of its temper, or ſuch, more probably, is its diſlike to the chill of a watery ſoil, that during the autumnal rains, which fall in the Deſarts of Zahara, where it chiefly inhabits, no traces of its lying down have ever yet been ſeen. In the ſtillneſs of the night it often ventures to the corn fields of Fezzan, where, in traps prepared for the purpofe, it is fometimes taken.

The *food* of the lower claffes of the people confifts of the flour of Indian corn, feafoned with oil; of dates, apricots, and pomegranates, and of calabafhes, cucumbers, and garden roots.

Perfons of a fuperior rank are alfo fupplied with wheat bread, which is baked in their own houfes; with mutton, goats flefh, the flefh of the camel, and that of the antelope; and with a great variety of fruits, and of garden vegetables.

Fezzan produces a fufficiency of falt for the confumption of its own inhabitants.

The water in general has a mineral tafte; yet fome of the fprings are pure: but

but the favourite beverage confifts of a liquor which the date tree, like the palm, affords. At firft it poffeffes the mild flavour and cooling quality of orgeat; but acquires, when fermented, an acefcent tafte and intoxicating ftrength that are ftill more highly valued.

To the palm the lofs of fo large a proportion of its fap is generally fatal; but the hardier date tree recovers from its wound, and in the courfe of two or three years regains its former health.

In their *Religion* the people of Fezzan are rigid Mahometans; not intolerant to the opinions of others, but ftrict and fuperftitious in the obfervance of their own.

The *Government* of Fezzan is purely monarchical; but its powers, which feem to be reſtricted by the influence of opinion, are adminiſtered with ſuch paternal regard to the happineſs of the people, the rights of property are ſo much revered, the taxes are ſo moderate, and the courſe of juſtice is directed by ſo firm, and yet ſo temperate a hand, that the inhabitants of Fezzan (as far as judgement can be formed from the feelings of thoſe who travelled with Mr. LUCAS) are affectionately and ardently attached to their Sovereign.*

The

* *To theſe ſentiments of conſtant regard and of deep veneration for their King, his acknowledged deſcent from the Prophet has undoubtedly contributed: for ſuch, if united to the temporal power, is the effect of this claim to religious authority, that in Morocco, when in the hour of his wrath the ſword of the Emperor is drawn, the ſubmiſſive victims whom chance or official attendance on the Court preſents to his fury, ſtretch forth their necks with ſilent and humble acquieſcence; perfectly convinced that the ſtroke of death, when given by ſo ſacred a hand, is an inſtant admiſſion to Paradiſe.*

The preſent King, Mohammed Ben Mohammed, is deſcended from one of the Shereefs of Tafilet, in the kingdom of Morocco, who was related to the Regal Family of that empire, and who, about 400 years ſince, obtained poſſeſſion of the Crown of Fezzan.

From that period to the middle of the preſent century, the kingdom maintained its Independence; but at the latter æra, the Baſhaw of Tripoli invaded it with a powerful force, laid ſiege to the capital, defeated, and took the King captive, and carried him a priſoner to Tripoli. For two years the unfortunate Monarch was detained in bondage, but at the end of that period, on the condition of an annual tribute of fifty ſlaves and ten pounds of gold duſt, the Baſhaw reſtored him to his Crown.

Till

Till the acceſſion of the preſent King of Fezzan, this tribute was faithfully tranſmitted. But the reigning Sovereign, conſcious of the declining power of Tripoli, and of the internal ſtrength which the affection and confidence of his people had given to his Kingdom, has gradually diminiſhed the amount, and altered the nature of the acknowledgement; for it now conſiſts in an occaſional preſent of a few ſlaves, and of a pound or two of gold duſt, and is rather the compliment of a trading State to the Kingdom in which its principal commodities are ſold, than a proof of political dependence. Still, however, the expreſſion of *my* Kingdom of Fezzan is in frequent uſe with the reigning Baſhaw of Tripoli, who is the grandſon of the Conqueror; nor has the dependence of the

the former State on the latter been ever directly renounced.*

In Fezzan, as in all the Countries in which the Mahometan Faith is eſtabliſhed, the deſcendants of the Prophet are conſidered as a diſtinct and highly privileged order. Their property is ſacred; their perſons are inviolable; and while the colour of their turbans, and the revered title of Shereef, announce to the people the auguſt dignity which they bear, they derive from the hereditary nature of their privileges a high degree of permanent influence, and ſometimes of dangerous power.‡

Among

* When I was at Fezzan, about twenty years ago, the actual government was committed to an Alcaid, who received his annual appointment from the Baſhaw of Tripoli.
BEN ALLI.

‡ *In Morocco, as in Fezzan, the Founder of the reigning family was indebted for the diadem to the reſpect and reverence which the title of Shereef beſtows.*

Among the privileges of their rank may be numbered an exemption from certain puniſhments, and that ſort of general indulgence which the prevalent dread of ſhedding the blood of the Prophet unavoidably creates.

To theſe circumſtances of diſtinction it is owing that, like the Nobility of other States, they are ſometimes ſubjected to, and feel a particular apprehenſion of the penalty of diſhonour, eſpecially that of having duſt thrown upon their heads.

But great as the privileges of this high order unqueſtionably are, the individuals who compoſe it have no union or general concert with each other; for neither in conteſts for power, nor in reſiſtance to oppreſſion, do they ever act as a body.

Some of the claſs are devoted to the

indolence

indolence of a religious life; but a larger proportion is engaged in the active concerns and progreſſive purſuits of commerce: for, in general, the deſcendants of Mahomet, following the example of their great anceſtor, are either Princes or Merchants.

The *Revenue* of Fezzan is compoſed of a variety of branches, the principal of which are,

1ſt, A tax on towns and villages, at the rate of from 100 to 500 mitkals each. The number of towns and villages is eſtimated at 100; and the value of a mitkal at about ten ſhillings Engliſh: conſequently, the tax on each town and village is from 50l. to 250l. ſterling.

2d, A tax on every camel's load of goods (proviſions for the market excepted)

cepted) which enters the capital. The rate of this impoſt is one mitkal, or ten ſhillings Engliſh, per load.

3d, The fines that are inflicted as puniſhments for offences.

4th, The produce of ſuch lands as, on the death of the poſſeſſors, without heirs, eſcheat to the Crown.

5th, A ſmall tax upon all gardens and date trees.

To theſe ſeveral branches of revenue may be added, as another ſource of conſiderable income, the commerce in which the King is engaged; for he ſeems, eſpecially in the articles of trona and ſenna, to be the principal Merchant in the kingdom.

Gold duſt conſtitutes in Fezzan the chief medium of payment; and value in that medium is always expreſſed by weight.

Their common meaſures of weight, and conſequently of value, are,

The xarobe (or harrobe) the weight of which is four grains, and which expreſſes the value of thirty-three piaſtres of Tripoli, or 6d. ſterling.

The mitkal, the weight of which is eighty-one grains, and which expreſſes the value of 668$\frac{1}{4}$ Tripoli piaſtres, or 10s. 1$\frac{1}{2}$d.

A ſingle grain therefore is equal in value to 1$\frac{1}{2}$d.

In the choice of the firſt of theſe mea-
ſures,

ſures, the xarobe, they ſeem to have been influenced by the diſcovery, 1ſt, That four grains is the uſual weight of a hard and durable berry,* which is called habbat ell goreth; and 2dly, That half that weight, two grains, is the weight of another and moſt beautiful berry, which is brought from Nigritia, and to which, from its ſcarlet colour, and the black ſtroke that ornaments one of its ends, they have given the name of eyne-deeka, or the cock's eye. Theſe berries are therefore employed as the common weights for gold duſt in Fezzan; for the xarobe and the mitkal have only an ideal exiſtence.

But for the greater convenience of exchange,

* *From the appearance of the ſuppoſed berries, there is reaſon to ſuſpect that they belong to the claſs of leguminous plants, and are in fact two ſpecies of peaſe.*

change, the Merchants of Fezzan are generally furniſhed with ſmall papers of gold duſt, of different values, from that of two xarobes, or one ſhilling, to ſuch an amount as their buſineſs is likely to require.

If the value of the article to be purchaſed is leſs than two xarobes, the payment is generally made in corn or flour. Thus the ſmaller articles of proviſions, as eggs, onions, &c. are generally purchaſed by a proportionable value in corn.

TABLE.

One grain (in weight) of gold is equal to - - - -	$8\frac{1}{4}$ piaſtres of Tripoli, $1\frac{1}{2}$d. ſterling.
Four grains ditto are equal to	1 xarobe of Fezzan, 33 piaſtres of Tripoli, 6d. ſterling.
Eight grains ditto are equal to	2 xarobes of Fezzan, 66 piaſtres of Tripoli, 1s. ſterling.
Twenty xarobes, or eighty grains, are equal to - -	* 1 mitkal (an imaginary coin of Fezzan) 666 piaſtres of Tripoli, 10s. ſterling.

In

* In the neighbourhood of Tombuctou a gold mitgan is worth about 10s. 6d. ſterling.

BEN ALLI.

In this view of the currency of Fezzan, the ſmall fractions that would be requiſite to render it perfectly accurate are omitted.

A mitkal is 675 piaſtres of Tripoli, or a fraction more than 10s. 1½d. ſterling, and conſequently it exceeds, by a fraction, the amount of 20 xarobes.

The grains of Fezzan are of the ſame * weight as in England, but the okea, or ounce of Fezzan, is very different, for it contains 640 grains; whereas the Engliſh

* Mr. LUCAS *found by his medicine ſcales, that the pea called habbat ellgoreth, which is uſed in Fezzan for a weight of four grains, is exactly equal to four grains Engliſh.*

In England one grain of gold is worth 2d.—one pennyweight is worth 4s.—and one ounce is worth 4l. ſterling.

REES's Edition of Chambers's Dictionary, Article "GOLD."

ounce contains but 480, which is a fourh lefs.

A Fezzan ounce of gold therefore, or 640 grains, at 1½d. per grain, muft be worth in Fezzan 4l.

And an Englifh ounce of gold, or 480 grains, at 1½d. per grain, muft be worth in Fezzan 3l.

Among the circumftances for which the natives of Fezzan, who travelled with Mr. LUCAS, confidered their Sovereign as eminently diftinguifhed, they often mentioned his juft and impartial, but fevere and determined adminiftration of juftice; and as a proof of the afcendancy which, in this refpect, he has acquired over the minds of his fubjects, they defcribed the following cuftom:—If any man has injured another, and refufes fuſes to go with him to the Judge, the complainant, drawing a circle round the oppreſſor, ſolemnly charges him in the King's name, not to leave the place till the Officers of Juſtice, in ſearch of whom he is going, ſhall arrive: and ſuch (if they are to be credited) is, on the one hand, his fear of the puniſhment which is inflicted on thoſe who diſobey the injunction, and ſo great, on the other, is his dread of the perpetual baniſhment which, if he ſeeks his ſafety by withdrawing from the kingdom, muſt be his inevitable lot, that this imaginary priſon operates as a real confinement, and the offender ſubmiſſively waits the arrival of the Officers of the Judge.

Small offences are puniſhed by the baſtinado: but thoſe of a greater magnitude ſubject the convict, according to the different degrees of guilt, to the penalty

nalty of a fine, of imprifonment, or of death.——Shereefs, like the Nobility of other States, are fometimes punifhed, as was mentioned before, by the pain of difhonour; in which clafs of punifhments, the moft dreaded, becaufe the moft reproachful, is the indignity of having duft heaped upon their heads.

To their infulated and remote fituation, and to their natural barrier of defolate mountains and dreary waftes of fand, much more than to military ftrength, the people of Fezzan are indebted for their fecurity.

Trufting to this natural defence, their towns are without guards, and, their capital excepted, are without walls; nor have they any regular ftanding force; yet the Shereef conceives that 15 or 20,000 troops might, upon an emergen-

cy,

cy, be raiſed. The only expedition of a military nature that has happened within his remembrance, was undertaken on the following account:—

South-Eaſt of the capital, at the diſtance of 150 miles, is a wide and ſandy deſart, entirely barren, and oppreſſed with a ſuffocating heat. Immediately beyond this deſart, the width of which is about 200 miles, the mountains of Tibeſti, inhabited by a wild and ſavage people of that name, begin to take their riſe. Ferocious in their manners, free-booters in their principles, and ſecure, as they thought, in the natural defences of their ſituation, theſe independent mountaineers became the terror of the caravans which traded from Fezzan to Bornou, and which are obliged to paſs the Weſtern extremity of the Deſart. But at length, having plundered a caravan which belonged to the King himſelf, and having killed about twenty of his people, their conduct provoked his reſentment, and determined him to revenge the inſult. With this view he immediately raiſed a ſmall army of from 3 to 4,000 men, the command of which he gave to an able and active *Magiſtrate*, announcing, by that appointment, that he ſent them, not to ſubdue a reſpectable enemy, but to puniſh an aſſemblage of plunderers and aſſaſſins. Having compleated the difficult paſſage of the deſart, and having gained the firſt aſcent of the mountains, they proceeded without oppoſition, till at length the natives, who waited in ambuſh, ruſhed upon them, and with the bows and arrows, and lances, with which they were armed, began a furious aſſault: but the inſtant that the foremoſt of the ſoldiers had given their fire, the mountaineers, more alarmed

alarmed at the dreadful ſounds which they heard, and at the imagined lightning which they ſaw, than terrified with the ſlaughter that was made, threw down their arms, and flying with great precipitation, abandoned, to the mercy of the victors, their houſes and their helpleſs inhabitants. The next morning, a deputation from the natives, of their principal people arrived at the camp, with humble intreaties that their wives and children might be ſpared, and an offer, on that condition, to ſubmit to any terms which the Alcaid ſhould deſire to impoſe. The Alcaid accordingly demanded, and received, as hoſtages for their future conduct, twenty of their principal people, with whom, and with all the plunder which the country afforded, he returned in triumph to Fezzan. There the King entertained them with kindneſs, and under a promiſe that their nation

tion ſhould acknowledge him as their *Sovereign*, and ſhould annually pay to him a tribute of twenty camel loads of ſenna, made them valuable preſents; and with ſtrong impreſſions on their minds, of the generous treatment which they had received, ſent them back to Tibeſti.

From that period no attempt to moleſt his caravans has been made by the mountaineers; and though they neither acknowledge the King of Fezzan for their Sovereign, nor pay him any tribute, yet they bring the whole of their ſenna to Mourzouk for ſale, where it is purchaſed to great advantage by the King, and is afterwards ſold, on his account, at the market of Tripoli.

An occaſional viſit to the Court of Fezzan is paid by their Chief, who is always received with great hoſpitality; and after a rea reſidence of a few weeks, is diſmiſſed, with a preſent of a long robe.

The vales of Tibeſti are fertile in corn, and paſturage for cattle, of which they have numerous herds, and are particularly celebrated for their breed of camels, which are eſteemed the beſt in Africa. For this fertility they are indebted to the water of the innumerable ſprings that amply compenſate for the want of rain, which ſeldom, if ever, falls within the limits of Tibeſti.

Huts of the ſimpleſt conſtruction (for they are formed of ſtakes driven into the ground in a circular arrangement, and covered with the branches of trees and bruſhwood intermixed) compoſe the dwellings of the people.

In return for the ſenna and the camels which

which they ſell in Fezzan, they bring back coral, alhaiks, or barakans, Imperial dollars, and braſs, from the two laſt of which articles they manufacture the rings and bracelets which are worn by their women.

Among the natives of Tibeſti different religions are profeſſed; for ſome of them are Mahometans, and others continue attached to their antient ſyſtem of Idolatry.

CHAPTER

CHAPTER V.

Mode of Travelling in Africa.

THE mode of travelling in Africa is fo connected with the commerce, and therefore with the manners of its principal nations, that without fome knowledge of the *former*, a defcription of the two *latter* cannot be clearly underftood.

In that divifion of Africa which lies to the North of the Niger, the feafon for travelling begins with the month of October, and terminates with the month of March. During this period, the tempe-

rature

rature of the air, though ſtrongly affect-ed by the degree of latitude, the eleva-tion of the land, the diſtance of the ſea, and the direction of the wind, is compa-ratively cool; and in ſome places, as in the neighbourhood of Mount Atlas, and on the Coaſts of the Mediterranean, oc-caſionally exhibits the phænomena of an European Winter. At Tripoli, the Thermometer is ſometimes ſeen at the 40th degree of Farenheit's ſcale; and on the 31ſt of December, in the year 1788, it was obſerved, at nine in the morning, to have fallen within four degrees of the freezing point; a coldneſs that was fol-lowed by a ſlight ſhower of ſnow.

In all countries, the animals which Nature and the attention of man have provided for the conveyance of the Tra-veller, and for the tranſit of his mer-chandize, are ſuited to the character of the

the ſoil, and to the ſmoothneſs or inequality of its ſurface. Of the ſoil of Africa, to the North of the Niger, the prevailing character is ſand; and though in the neighbourhood of rivers, and in all thoſe diſtricts which receive from the adjacent mountains, the advantage of numerous ſprings, the ſand is blended with a vegetable mould, yet the ground, in general, is remarkably ſoft and dry. In general, too, the ſurface of the land, though in ſome places broken by naked rocks, and ſwelling, in others, to mountains of conſiderable magnitude, may be regarded as comparatively level.

To ſuch a country the camel is peculiarly ſuited; for his broad and tender foot, which ſlides on a wet ſurface, and is injured by the reſiſtance of ſtones, is obſerved to tread with perfect ſecurity and eaſe on the dry and yielding ſand:

and

and while, from the ſame circumſtances in its ſtructure, his hoof is incapable of faſtening, with any ſtrength, on the ground of a ſteep aſcent, and furniſhes, in a ſhelving declivity, no ſolid or ſufficient ſupport, his movement on a ſmooth and level ſurface is ſingularly firm and ſafe.

So remarkably exemplified in Africa is that rule in the œconomy of Nature which ſuits the beaſt of burden to the land which it inhabits, that in the country which lies to the South-Weſt of the Niger, where the ſurface is mountainous, and the ground is as ſtony in ſome place, as it is wet and muddy in others, no camels are found. Their place is ſupplied by ſmall horſes, aſſes, and mules.

The proper burthen for a camel varies with its ſtrength, which is very different

ferent in different ſpecies of the animal. In the dominions of Tripoli, a common load is from three to four hundred weight; and the medium expence of the conveyance for each hundred appears to be one farthing per mile.

The uſual rate of travelling is three miles in the hour, and the number of hours that are actually employed on the rout, excluſive of thoſe which are allotted to refreſhment, is ſeldom more than ſeven or eight in a day. Of the number of days which are conſumed on a long journey, many are devoted to the purpoſe of occaſional trade, to that of recruiting the ſtrength of the camels, and to that of procuring additional ſtores of proviſions and of water; for in all ſuch places as are able to furniſh a ſupply of proviſions (which are generally places of conſiderable population, and

 there

therefore of ſome traffic) the ſtay of the caravan is ſeldom leſs than two days, and is often prolonged to more.

The general food of the camels is ſuch only as their nightly paſture affords; and is often confined to the hard and thorny ſhrubs of the Deſart, where a ſullen vegetation is created by the rains of the Winter, and upheld by the dews that deſcend in copious abundance through all the remainder of the year.

Of the drivers and ſervants of the caravan, the cuſtomary food conſiſts of the milk of the camels, with a few dates, together with the meal of barley or Indian corn, which is ſometimes ſeaſoned with oil, while the Merchant ſuperadds, for his own uſe, the dried fleſh of the camel, or of ſheep, and concludes his repaſt with coffee.

Water

Water is drawn from the wells in leather buckets, that form a part of the travelling equipage of the caravan, and is carried in the ſkins of goats, through which, however, though tarred within and without, it is often exhaled by the heat of the noon-day ſun.

A particular mode of eaſy conveyance is provided for the women and children, and for perſons oppreſſed with infirmity or illneſs. Six or eight camels are yoked together in a row, and a number of tent poles are placed in parallel lines upon their backs: theſe are covered with carpets; and bags of corn are ſuperadded to bring the floor to a level, as well as to ſoften the harſhneſs of the camel's movement: other carpets are then ſpread, and the traveller ſits or lies down, with as much convenience as if he reſted on a couch.

The Desart (a term of the same meaning with its Arabic name of Zahara) may be said, like the ocean, to connect the very nations which it seems to separate; for, in comparison with the woods and morasses of America, it furnishes the Merchant with an easy and convenient road.

A sandy heath of various levels, in some places naked and bare, but much more frequently covered with an odoriferous plant, which the Arabs call the Shé, and which, though far superior in fragrance, has at least a remote resemblance to the wild thyme of Europe, constitutes the general appearance of the Desart. The exceptions, however, are interesting and important: for besides the diversity that arises from the different shrubs, which are often scantily intermixed with the Shé, and of which the thorny

thorny plant that forms the harſh food of the camel appears to be the moſt common, an eſſential variation is furniſhed by the comparative fertility of ſome particular diſtricts, and by the total barrenneſs of others.

In ſome portions of the general wilderneſs, thouſands of ſheep, and goats, and cows, are ſeen to paſture; while in others nothing is preſented to the eye but deſolate hills of ſhifting ſand.

To the laſt of theſe the name of *Deſarts without Water* is emphatically given; a name that conveys to an Arab ear the fearful idea of an intenſe and ſuffocating heat, of the total abſence of vegetable life, and of the hazard of a dreadful death. For though the movement of the ſhifting ſands is not ſo rapid as to endanger the ſafety of the caravan, yet as the ſcorching heat of the ſun-beams, confined and reflected by the hills of ſand, hourly diminiſhes the ſtore of water, and as the breadth of ſeveral of theſe deſarts is that of a ten days journey, the hazard of periſhing with thirſt is ſometimes fatally experienced.

All means of aſcertaining the rout by land-marks, the uſual guides in other parts of the wilderneſs, are here deſtroyed by the varying forms and ſhifting poſitions of the hills: but from anxious obſervation and continued practice, the camel-drivers acquire a ſufficient knowledge of the bearings of the ſun and ſtars to direct them in their way.

Such are the expedients by which the difficulties of the Deſart are in general overcome: thoſe which are preſented by the broad current of the impetuous Niger,

ger, though much more eaſily, are not ſo frequently ſurmounted.

Of this river, which in Arabic is ſometimes called Neel il Kibeer, or the Great Nile, and at others, Neel il Abeed, or the Nile of the Negros, the riſe and termination are unknown, but the courſe is from Eaſt to Weſt. So great is its rapidity, that no veſſel can aſcend its ſtream: and ſuch is the want of ſkill, or ſuch the abſence of commercial inducements among the nations who inhabit its borders, that even *with* the current, neither veſſels nor boats are ſeen to navigate. In one place, indeed, the Traveller finds accommodations for the paſſage of himſelf and of his goods; but even there, though the ferrymen, by the indulgence of the Sultan of Caſhna, are exempted from all taxes, the boat which conveys the merchandize is nothing more than an ill-conſtructed raft; for the planks are faſtened to the timbers with ropes, and the ſeams are cloſed both within and without, by a plaiſter of tough clay, of which a large proviſion is always carried on the raft for the purpoſe of excluding the ſtream wherever its entrance is obſerved.

The depth of the river at the place of paſſage, which is more than a hundred miles to the South of the City of Caſhna, the capital of the empire of that name, is eſtimated at twenty-three or twenty-four feet Engliſh. *

Its width is ſuch that even at the Iſland of Gongoo, where the ferrymen reſide,

* Its depth is from ten to twelve peeks, each of which is twenty-ſeven inches.

The Shereef IMHAMMED.

ſide, the ſound of the loudeſt voice from the northern ſhore is ſcarcely heard; and at Tombuctou, where the name of Gnewa, or black, is given to the ſtream, the width is deſcribed as being that of the Thames at Weſtminſter. In the rainy ſeaſon it ſwells above its banks, and not only floods the adjacent lands, but often ſweeps before it the cattle and cottages of the ſhort-ſighted or too confident inhabitants.

That the people who live in the neighbourhood of the Niger ſhould refuſe to profit by its navigation, may juſtly ſurpriſe the Traveller: but much greater is his aſtoniſhment, when he finds that even the food which the bounty of the ſtream would give, is uſeleſsly offered to their acceptance; for ſuch is the want of ſkill, or ſuch the ſettled diſlike of the people to to this ſort of proviſion, that the fiſh with which the river abounds are left in undiſturbed poſſeſſion of its waters.

Having paſſed the ſtream, the face of the country, and with it the mode of travelling, are changed. High mountains and narrow valleys, extenſive woods and miry roads, ſucceed to the vaſt plains and ſandy ſoil of the Zahara and its neighbouring kingdoms. Water is no longer refuſed or ſcantily given to the parched lips of the Traveller; but while the abundance of this refreſhment, and of the vaſt variety both of vegetable and animal food that is offered in profuſion for his ſupport, diminiſh the hardſhips and remove the principal hazards of his journey, the raging heat of the Torrid Clime increaſes as he proceeds. Wet cloths applied to the mouth are ſome-

times

times requiſite, and eſpecially in the woods, to allay, for the purpoſe of reſpiration, the violence of the burning air; and the journey, which the fierceneſs of the ſun ſuſpends, is often renewed amidſt the dews and comparative darkneſs of the night.

From the want of camels, which are ſeldom ſeen to the South of Caſhna, nor even to the Weſt, except in higher latitudes, the conveyance of the Merchant and his goods is committed to the mules, and ſmall horſes and aſſes of the country. Of the firſt, the uſual burden is 200lbs. of the ſecond, 150lbs. and of the laſt, 100lbs.

Travelling through all this part of Africa is conſidered as ſo ſecure, that the Shereef Imhammed, with the utmoſt chear-

chearfulneſs and confidence of ſafety, propoſed to accompany and conduct Mr. LUCAS, by the way of Fezzan and Caſhna, acroſs the Niger, to Aſſenté, which borders on the Coaſt of the Chriſtians.

CHAPTER VI.

General Remarks on the Empires of Bornou and Caſhna—Rout from Mourzouk to Bornou—Climate of Bornou—Complexion, Dreſs, and Food of the Inhabitants——Their Mode of Building—Their Language—Government—Military Force—Manners—and Trade.

TO the South of the Kingdom of Fezzan, in that vaſt region which ſpreads itſelf from the river of the Antelopes weſtward for 1200 miles, and includes a conſiderable part of the Niger's courſe, two great empires, thoſe of Bornou and thoſe of Caſhna, are eſtabliſhed.

The

The circumſtances of ſoil and climate, and thoſe alſo which conſtitute political character, are nearly the ſame in both: for their prevailing winds are the ſame; their rains, which are periodical in each, though much more profuſe in Bornou, begin at the ſame ſeaſon; the ſame grains are cultivated; the ſame fruits (generally ſpeaking) are produced; and except that no camels are bred to the weſtward of the City of Caſhna, the capital of the empire, the ſame animals are reared. In both, the complexion of the inhabitants is black; their mode of building too is ſimilar, and their manners, though in ſome reſpects more civilized in Bornou, have a general reſemblance.

Each of the two empires is formed by the ſubjection of different tribes or nations to the dominion of one ruling people.

ple. The nature of the Government, and the laws which regulate its ſucceſ-ſion, are the ſame in both. In both, the ruling people are Mahometans; in both, the dependent nations are compoſed of converts to the Muſſelman faith, and of adherents to the antient worſhip; and though at preſent their languages are different, the conquerors in both had probably the ſame original.

Of theſe empires, Caſhna, till of late, was eſteemed the firſt in power; but though a thouſand villages and towns are ſtill included in her vaſt domains, ſhe is now conſidered as much inferior to Bornou.

ROUT

ROUT FROM MOURZOUK TO BORNOU.

FROM Mourzouk in Fezzan to Bornou, the capital of the empire from which it takes its name, the Fezzaners, whoſe commercial ſpirit no diſtance can diſcourage, are conducted by a rout of more than a thouſand miles. Temiſſa, the firſt town at which they arrive, and the laſt which they ſee in Fezzan, they reach on the ſeventh day; and in three days more they enter the territories of Bornou. Several villages, inhabited by Blacks, whoſe perſons, their waiſts excepted, are entirely naked, whoſe meagre limbs and famiſhed looks announce the extremity of miſery, and whoſe idolatrous religion neither excites the reſentment nor reſtrains the charity of the benevo-

benevolent Fezzaners, mark the northern frontier of the empire.

On the day following their departure from theſe melancholy hamlets, they begin the aſcent of a hilly uninhabited deſart of ſand, where a few buſhes of penurious vegetation point out the ſucceſſive wells that are found in theſe barren heights, and diminiſh the fatigues of a three days paſſage. At the cloſe of the fourth day they enter a plain that is inhabited by Mahometans, where, in addition to a plentiful ſupply of excellent water, they are cheared with the ſight of date trees, and of Indian corn.

From this plain, which lies to the Weſt of the Deſart of Tibeſti, and the end of which they reach on the ſecond day, a part of the Tibeſti mountains take their riſe. Theſe vaſt hills, the range of which is very extenſive, are variouſly peopled: but ſuch of them as are croſſed on the rout from Fezzan to the City of Bornou, are inhabited by a mixture of Muſſelmen and Idolaters, who employ themſelves in breeding camels and aſſes, and other cattle, particularly horſes of a ſmall ſize.

Excluſively of the two days that are requiſite for the paſſage of the mountains, an allowance of twice that time is generally given to refreſhment and repoſe; ſoon after which a fertile and beautiful country, as richly diverſified as numerouſly peopled, opens to their view. Its inhabitants are herdſmen, and with the exception of a few Pagans who are intermixed among them, are Muſſelmen in their faith. Their dwellings are in tents

tents, which are compoſed of hides, and their wealth conſiſts in the multitude of their cows and ſheep.

Four days are employed in croſſing theſe fortunate diſtricts: the ſixth conducts the caravan to the entrance of the vaſt and burning Deſart of Bilma. Surrounded by this dreary ſolitude, the Traveller ſees with a dejected eye the dead bodies of the birds that the violence of the wind has brought from happier regions; and as he ruminates on the fearful length of his remaining paſſage, liſtens with horror to the voice of the driving blaſt, the only ſound that interrupts the awful repoſe of the Deſart.

On the eleventh day from their entrance on theſe ſcorching ſands, the caravan arrives in the fertile plains that encompaſs the Town of Domboo, the approach to which is enlivened by the frequent appearance of the majeſtic Oſtrich, and of the gay but fearful Antelope.

From thence, in about five days, they reach the City of Kanem, the capital of an extenſive and fertile province, of which it bears the name, and in which the inhabitants, who are compoſed of Muſſelmen and Pagans, breed multitudes of cattle, and raiſe innumerable horſes for the ſervice of the King.

A journey of ten days more concludes their labour, and brings them to the imperial City of Bornou.*

Bornou,

* The rout which BEN ALLI purſued from Fezzan to Bornou is not diſtinctly deſcribed.

His

Bornou, the name which the natives give to the country, is diſtinguiſhed in

 Arabic

His relation is, that on the 26th day from the time of his leaving Fezzan, he arrived at a place which in Arabic is called Wéddan, or the Rivers, for Weddan is the plural of Wed, which ſignifies a river.

The firſt part of the country through which he paſſed is repreſented as a ſandy Deſart, in which the Shé (a plant that reſembles the Wild Thyme of England) and a few buſhes of ſhrubs and ſhort trees are thinly ſcattered, and wells of water are extremely rare. Wandering Arabs, of the powerful but hoſpitable Tribes of Booaiſh and Duhaſſin, appear to be its only inhabitants; and Wéddan itſelf is ſaid to contain but 130 houſes, which are built of earth and ſand; and to furniſh no articles of trade but dates and ſalt; yet the country around it is called prolific: the rice grounds are deſcribed as numerous; and multitudes of ſheep and goats, of camels and of horſes, ſwell the liſt of its poſſeſſions.

He repreſents the Duhaſſin Arabs, as Merchants journeying to Bornou, who carried with them for ſale an aſſortment

Arabic by the appellation of Bernou or Bernoa, a word that ſignifies the land of

---

ment of goods; among which he enumerates wheat, barley, dates, ſalt, tobacco, and alhaiks: and he obſerves, that he purchaſed from the Chief the permiſſion of accompanying the Tribe, and the conſequent aſſurance of a ſafe paſſage to Bornou.

From Wéddan, by forced marches, they arrrived in twenty days at Bornou. A deſart of ſand, in ſome places interrupted by woods, and occaſionally watered with rivulets of a ſtrong mineral taſte, conſtitutes the general deſcription of the country. But as he entered the kingdom of the Bornoos, the limit of which he repreſents as ſeven days diſtant from the capital, he paſſed through ſeveral poor villages of Blacks, who live upon the charity of Travellers; for though there be no regular marked road, yet the caravans always take the ſame rout, and paſs by thoſe villages both in going and returning.

BEN ALLI ſeems to have travelled from Mourzouk to Bornou by a different rout from that which is uſually taken by the Merchants of Fezzan: nor can it be ſuppoſed, that the independent and powerful Arabs with whom he journeyed, of Noah, for the Arabs conceive that, on the first retiring of the deluge, its mountains received the Ark.

The *Climate*, as may naturally be expected in a kingdom which seems to be bounded by the 16th and the 26th parallels of latitude, is characterized by excessive, though not by uniform heat.

neyed, would either obtain, or solicit the permission of the Sovereign of Fezzan to pass in so large a body through his small and unguarded dominions. And though the corresponding accounts that are given in the narratives of the Shereef and of BEN ALLI, of some villages of miserable Blacks, may suggest an idea that two roads intersected each other on the frontier of Bornou, yet as on that supposition, the different times within which the several parts of the two journeys were respectively performed cannot be easily reconciled, there is reason to believe that the villages described by BEN ALLI, though peopled by similar inhabitants, may not be the same with those which attracted the compassionate notice of the Shereef.

Two ſeaſons, the one commencing ſoon after the middle of April, the other at the ſame period of October, may be ſaid to divide the year. The *firſt* is introduced by violent winds that bring with them, from the South Eaſt and South, an intenſe heat, with a deluge of ſultry rain, and ſuch tempeſts of thunder and lightning as deſtroy multitudes of the cattle, and many of the people. During the rainy period (the continuance of which is from three to nine ſucceſſive days, with ſhort intervals from the occaſional changes of the wind to the North or Weſt) the inhabitants confine themſelves cloſely to their dwellings; but the reſt of the firſt ſeaſon, however ſultry and however occaſionally wet, is not incompatible with the neceſſary labours of the huſbandman and the ſhepherd.

At the commencement of the *ſecond* ſeaſon

ſeaſon in the latter part of October, the ardent heat ſubſides; the air becomes ſoft and mild; the weather continues perfectly ſerene; and as the year declines, an unwelcome coolneſs precedes the riſing of the ſun.

The inhabitants, though conſiſting of ſuch a multitude of nations that thirty languages are ſaid to be ſpoken in the empire, are alike in their *Complexion*, which is entirely black; but they are not of the Negro caſt.

In a climate ſo warm, the chief recommendations of *Dreſs* are decency and ornament: among the pooreſt, therefore, by whom the firſt only is regarded, a kind of girdle for the waiſt is ſometimes the only covering; but in general a turban, conſiſting, as in Barbary, of a red woollen

len cap, ſurrounded by folds of cotton, together with a looſe robe of coloured cotton of a coarſer kind, are alſo worn.*

The *Grain* that conſtitutes the principal object of culture in Bornou is Indian corn, of two different kinds, which are diſtinguiſhed in this country by the names of the gaſſŏb, and the gamphûly.

The gaſſŏb, which in its general ſhape reſembles the common reed, is of two ſpecies. The firſt grows with a long ſtalk that

* The dreſs of the greateſt part of the people is compoſed of ſhirts of blue cotton, which is manufactured in the country; of a red cap, which is imported from Tripoli; and of a white muſlin turban, which is brought from Cairo by the pilgrims who return through that City from Mecca. Noſe-rings of gold are worn by the principal people as a mark of diſtinction.

BEN ALLI.

that bears an ear, which in length is from eight to twelve inches, and contains, in little huſks or cavities, from three to five hundred grains, of the ſize of ſmall peaſe. The ſecond ſpecies, which is common in Tripoli, differs no otherwiſe from the firſt than in the ſhorter ſize of the ear.

The gamphûly is diſtinguiſhed from the gaſſōb, by the bulk of the ſtalk; for that of the gamphûly is much thicker, by the number of its ears, for it has ſeveral on the ſame reed, and by the ſize of the grain, which is conſiderably larger. This kind of corn is frequently ſeen in Spain, and is there called Maize.

Wheat and barley are not raiſed in Bornou; but the horſe-bean of Europe and the common kidney-bean are cultivated with great aſſiduity, as they are uſed ufed for food, both by the flaves and by the cattle.*

In the culture of thefe different grains, the hoe alone is employed, as the ufe of the plough is ftill unknown to the people. The women divide with the men the labours of their hufbandry; for while the latter, with their hoes, open the ground, and form the trenches in ftraight lines parallel to each other, the women follow and throw in the feed: nor is this the only part which they take in the bufinefs of the field; for to them, as

* The country in the neighbourhood of the City of Bornou is fertile in Indian corn and rice. Of barley and wheat the quantity raifed is fmall. A fpecies of bean, which refembles the horfe-bean of Europe, though larger, and of a darker hue, is a much more common produce. Gum-trees are thinly fcattered. Cotton, hemp, and indigo, are alfo among the various produce of its foil.

BEN ALLI.

as ſoon as the weeds begin to riſe on the ridges of the lines in which the grain is ſowed, the hoe is conſtantly transferred.

The ſowing ſeaſon commences at the end of the periodical rains of April; and ſuch in that climate is the rapid vegetation, that on the 9th of July the gaſſôb is reaped; but the gamphúly, a grain of ſlower growth, is ſeldom cut till the month of Auguſt or September.

Such are the ſeveral ſpecies of corn that, among the people of Bornou, ſupply the place of the wheat, the barley, and the oats of Europe. Two ſpecies of roots are alſo uſed as wholeſome and ſubſtantial food: the one, which is called the Dondoo, produces a low plant, with branches that ſpread four or five feet upon the ground, and leaves that reſemble thoſe of the garden-bean. At the end

end of five months, from the time of its being planted, the leaves fall off, and the root is taken from the ground, and being cut into ſmall pieces, is dried in the ſun, in which ſtate it may be kept for two years. Its farther preparation conſiſts in reducing it to a fine powder, and mixing it with palm oil till it aſſumes the conſiſtency of paſte.

The other root is that of a tree, of which the name had eſcaped the Shereef's recollection: boiling is the only proceſs that is requiſite in preparing it for uſe.

The ſame character of ſufficiency which marks the catalogue of the different kinds of grain in Bornou, belongs alſo to the liſt of its various *Fruits*; for though neither olives nor oranges are ſeen in the empire, and even figs are rare; and

and though the apples and plumbs of its growth deſerve no commendation, and the dates are as indifferent as they are ſcarce, yet grapes, and apricots, and pomegranates, together with lemons and limes, and the two ſpecies of melons, the water and the muſk, are produced in large abundance.* But one of the moſt valuable of its vegetable ſtores, is a tree which is called Kedéynah, that in form and height reſembles the Olive, is like the Lemon in its leaf, and bears a nut, of which the kernel and the ſhell are both in great eſtimation, the firſt as a fruit, the laſt on account of the oil which it furniſhes when bruiſed, and which ſupplies the lamps of the people of Bornou with a ſubſtitute for the oil of olives.

To

* The country abounds in different ſpecies of fruit trees, but that which produces the date is not of the number.

BEN ALLI.

To this competent provifion of fuch vegetables as are requifite to the fupport, or grateful to the appetite of man, muft be added a much more ample and more varied fupply of *Animal Food*. Innumerable flocks of fheep, and herds of goats and cows, (for there are no oxen) together with multitudes of horfes, buffaloes, and camels, (the flefh of which is in high eftimation) cover the vales or pafture on the mountains of Bornou.*

The common, though not the Guinea fowl is alfo reared by the inhabitants; and their hives of bees are fo extremely numerous, that the wax is often thrown away as an article of no value in the market.

Their

* Horfes and horned cattle, goats and fheep, and camels, are the common animals of the country.

BEN ALLI.

Their *Game* conſiſts of the Huaddee, and other ſpecies of antelopes, of the partridge, the wild duck, and the oſtrich, the fleſh of which they prize above every other.

Their other wild animals are, the lion, the leopard, the civet cat, the ſmall wolf, the fox, the wild dog, that hunts the antelope; the elephant, which is not common, and of which they make no uſe; the crocodile, the hippopotamus, which is often killed on the banks of the river that runs from the Neel Shem, (the Nile of Egypt) to the Deſart of Bilma; and a large and ſingular animal, which is diſtiguiſhed by the name of Zarapah, and which is deſcribed as reſembling the camel in its head and body, as having a long and ſlender neck like the oſtrich, as being much taller at the ſhoulders than the haunches, and as defended by ſo tough a ſkin, as to furniſh the natives with ſhields that no arrow or javelin can pierce.*

Bornou, like other countries that approach the Equinoctial, is much infeſted with different kinds of dangerous or diſguſting reptiles, eſpecially ſnakes and ſcorpions, centipedes and toads.

Of its *beaſts of burthen* the variety is as ample as the numbers are abundant; for the camel, the horſe, the aſs, and the mule, are common in the empire.

The dog, with which the inhabitants purſue their game, appears to be their only *domeſtic* animal.

In

* *Giraffa is the name by which the Camelopardalis is called in the old zoological books.—The deſcription here inſerted, ſeems to have ariſen from a blended recollection of that animal; and of the hippopotamus, whoſe hide is extremely tough.*

In the mountains of Tibeſti, and perhaps in other parts of the empire, the herdſmen, probably for the ſake of a more eaſy change of paſture, prefer a reſidence in tents to ſtationary dwellings; and thoſe, it ſeems, are not manufactured, like the tents of the Zahara, from the camel's hair; but are compoſed of the hides of cows, a more durable and impervious covering.

Through all the empire of Bornou the ſame mode of *building*, and with the difference of a greater or a ſmaller ſcale, the ſame form in the plan of the houſes univerſally prevails.—Four walls, incloſing a ſquare, are erected: within thoſe walls, and parallel to them, four other walls are alſo built: the ground between the walls is then divided into different apartments, and is covered with a roof. Thus the ſpace within the interior walls determines the ſize of the court; the ſpace between the walls determines the width of the apartments; and the height of the walls determines the height of the rooms. In a large houſe the rooms are each about twenty feet in length, eleven feet in height, and as many in width.

On the outſide of the houſe, a ſecond ſquare or large yard, ſurrounded by a wall, is uſually provided for the incloſure and protection of the cattle.*

Such is the general plan of a Bornou houſe. For the conſtruction of the walls the following method is conſtantly purſued: a trench for the foundation being made, is filled with dry and ſolid materials rammed in with force, and levelled: on

* In form, the houſes are ſimilar to thoſe of Tripoli.
BEN ALLI.

on theſe a layer of tempered mud or clay is placed; and in this ſubſtitute for mortar a ſuitable number of ſtones are regularly fixed. Thus with alternate layers of clay and ſtones the wall proceeds; but as ſoon as it has reached the height of ſix or ſeven feet, the workmen ſuſpend its progreſs for a week, that it may have time to ſettle, and become compact; for which purpoſe they water it every day.

When the walls are finiſhed they are neatly plaiſtered, both within and without, with clay or mud, tempered with ſand; for the country furniſhes no lime.

The roofs are formed of branches of the palm tree, intermixed with bruſhwood; and are ſo conſtructed as at firſt to be water-proof; but ſuch is the violence of the wind and rain, that the end of the ſecond year is the utmoſt period of their duration.*

Much leſs attention is given to the furniture than is beſtowed on the ſtructure of the houſes; for the catalogue of utenſils is extremely ſhort. Among the lower claſſes of the inhabitants it conſiſts of the mats covered with a ſheep-ſkin, upon which they ſleep; of an earthen pot; of a pan of the ſame materials; of two or three wooden diſhes, a couple of wooden bowls, an old carpet, a lamp for oil, and perhaps a copper kettle.

Perſons

* The walls of the greateſt part of the houſes are built of a compoſition of earth and ſand, and are often waſhed down by the heavy rains; but others are formed of the more durable materials of ſtones and bricks.

The roofs are compoſed of the branches of trees, which are covered with layers of earth, and the whole building is white-waſhed with a ſpecies of chalk.

BEN ALLI.

Perſons of a ſuperior rank are alſo poſ- ſeſſed of leathern cuſhions, that are ſtuff- ed with wool; of ſeveral braſs and cop- per utenſils, of a handſome carpet, and of a ſort of candleſticks; for inſtead of the vegetable oil which is uſed by the common people, they employ the light of candles manufactured from their bees wax and the tallow of their ſheep.

Bornou is ſituated at the diſtance of a day's journey from a river which is call- ed Wed-el-Gazel, from the multitude of antelopes that feed upon its banks, and which is loſt in the deep and ſandy waſtes of the vaſt Deſart of Bilma.*

From the ſymmetry of the houſes,

P 4 and

* Bornoo is ſituated in a flat country, on the banks of a ſmall river.

BEN ALLI.

and the general reſemblance which they bear to each other, a regular arrangement of ſtreets might, with the utmoſt eaſe, have been given to their towns. In Bornou, however, a different ſyſtem has prevailed; for even in the capital, the houſes, ſtraggling wide of each other, are placed without method or rule; and the obvious propriety of giving to the principal moſque, a central ſituation, exhibits the only proof of attention to general convenience.*

The King's palace, ſurrounded by high walls, and forming a kind of citadel, is built,

* Bornoo, though a town of greater extent than Tripoli, conſiſts of a multitude of houſes, ſo irregularly placed that the ſpaces between them cannot be called ſtreets. It is furniſhed with moſques, which are conſtructed of brick and of earth; and with ſchools, in which the Koran is taught, as in the principal towns of Barbary.

BEN ALLI,

built, perhaps with a view to ſecurity, in a corner of the town.*

Markets for the ſale of proviſions are opened within the city; but for other articles, a weekly market, as in Barbary, is held without the walls.

	£.	s.	d.
The common price in Bornou of a cow or a bull is a mahaboob of Tripoli, or ...	0	6	0
A ſheep, ...	0	3	0
An oſtrich, ...	0	6	0
An antelope, ...	0	1	6
A camel from 6l. to 7l. 10s. or at a medium,	6	15	0
A horſe from 3l. to 7l. 10s. or at a medium,	5	5	0

In general, the towns have no other defence than that which the courage of the inhabitants affords; but the capital is

* In time of peace the Sultan always reſides in the capital.

BEN ALLI.

is ſurrounded by a wall of fourteen feet in height, the foundations of which are from eight to ten feet deep, and which ſeems to be built with conſiderable ſtrength. To this defence is given the additional ſecurity of a ditch, which encompaſſes the whole; and care is taken, that at ſun-ſet the ſeven gates which form the communication with the country ſhall be ſhut.*

The great population of Bornou is deſcribed by the indefinite and metaphorical expreſſion of a countleſs multitude.

In Fezzan the price of all things is meaſured by grains of gold; and where the value is too ſmall to be eaſily paid in

ſo

* Bornoo is ſurrounded by a wall, on which, however, there are no guards.

BEN ALLI.

ſo coſtly a metal the inhabitants have recourſe to corn, as a common medium of exchange. But in Bornou, as in Europe, the aid of inferior metals is employed, and copper and braſs (which ſeem to be melted together, and to be mixed with other materials) are formed into pieces of different weights, from an ounce to a pound, and conſtitute the current ſpecies of the empire.*

Dominions ſo extenſive as thoſe of Bornou have ſeldom the advantage of one uniform language; but an inſtance of ſo many different tongues, within the limits of one empire, as are ſpoken in that kingdom, and its dependencies, has ſtill

* A ſmall quantity of gold duſt is produced in Bornoo; but the principal medium of exchange conſiſts of pieces of metal which has ſome reſemblance to tin.

BEN ALLI.

ſtill leſs frequently occurred, for they are ſaid to be more than thirty in number.

Of the language, however, which is current in the capital, and which ſeems to be conſidered as the proper language* of Bornou, the following ſpecimen is given by the Shereef.

One is expreſſed by	*Lakka*	Eight is expreſſed by	*Tallóre*
Two	*Endee*	Nine	*L'ilkar*
Three	*Nieſkoo*	Ten	*Meiko*
Four	*Dekoo*	Eleven	*Meiko Lakka*
Five	*Okoo*	Twelve	*Meiko Endee*
Six	*Araſkoo*	Thirteen . . .	*Meiko Nieſkoo*
Seven	*Huſkoo*	Fourteen . . .	*Micko Deko*

Two

* The language of the common people of Bornoo, though different from, has a ſtrong reſemblance to that of the neighbouring Negros, and is very unlike the Arabic, in which, however, the Nobles and principal families converſe.

The

Two different *Religions* divide the ſentiments, without diſturbing the peace of the kingdom.

The ruling people profeſs the Mahometan faith; * and though the antient Paganiſm of the dependent nations does not appear to ſubject them to any inconvenience, a conſiderable part are converts to the doctrines of the Prophet.

An elective monarchy conſtitutes the *Government* of Bornou,† and like the ſimilar ſyſtem of Caſhna, endangers the happineſs,

The art of writing is known among them, and they are taught to expreſs the Bornoo tongue in the characters of the Arabic.

BEN ALLI.

* The Sultan and his ſubjects are Muſſelmen.

BEN ALLI.

† Bornoo is governed by a King, who takes the title of Sultan.

BEN ALLI.

pineſs, while it acknowledges the power of the people. On the death of the Sovereign, the privilege of chuſing among his ſons, without regard to priority of birth, a ſucceſſor to his throne, is conferred by the nation on three of the moſt diſtinguiſhed men, whoſe age and character for wiſdom, are denoted by their title of Elders; and whoſe conduct in the State has inveſted them with the public eſteem. Bound by no other rule as to their judgment, or reſtraint as to their will, than that which the expreſſed or implied inſtruction of electing the moſt worthy may form, they retire to the appointed place for their ſecret deliberation, the avenues to which are carefully guarded by the people: and while the contending ſuggeſtions of private intereſt, or a ſenſe of the real difficulty of chuſing where judgment may eaſily err, and error may be fatal to the State, keeps

keeps them in ſuſpence, the Princes are cloſely confined in ſeparate chambers of the Palace. Their choice being made, they proceed to the apartment of the Sovereign elect, and conduct him, in ſilence, to the gloomy place in which the unburied corpſe of his father, that cannot be interred till this awful ceremony is paſſed, awaits his arrival. There, the Elders point out to him the ſeveral virtues and the ſeveral defects which marked the character of his departed parent; and they alſo forcibly deſcribe, with juſt panegyric, or ſevere condemnation, the ſeveral meaſures which raiſed or depreſſed the glory of his reign. "You "ſee before you the end of your *mortal* "career: the *eternal*, which ſucceeds to "it, will be miſerable or happy in pro-"portion as your reign ſhall have proved "a curſe or a bleſſing to your people."

From

From this dread ſcene of inſtruction, the new Sovereign, amidſt the loud acclamations of the people, is conducted back to the Palace, and is there inveſted by the electors with all the ſlaves, and with two-thirds of all the lands and cattle of his father; the remaining third being always detained as a proviſion for the other children of the deceaſed Monarch. No ſooner is the Sovereign inveſted with the enſigns of Royalty, than ſuch of his brothers as have reached the age of manhood proſtrate themſelves at his feet, and in riſing preſs his hands to their lips—the two ceremonies that conſtitute the declaration of allegiance.

If any doubt of their ſincerity ſuggeſts itſelf to the King, or to the Elders, death or perpetual impriſonment removes the fear; but if no ſuſpicion ariſes, an eſtabliſhment of lands and cattle from the poſſeſſions of their father, together with preſents of ſlaves from the reigning monarch, are liberally beſtowed upon them.

Often, however, the moſt popular, or the moſt ambitious of the rejected Princes, covering his deſigns with cloſe diſſimulation, and the zeal of ſeeming attachment, creates a powerful party; and aſſured of foreign aid, prepares, in ſecret, the means of ſucceſsful revolt. But, ſtained with ſuch kindred blood, the ſceptre of the victorious Rebel is not laſtingly ſecure—one revolution invites and facilitates another; and till the ſlaughter of the field, the ſword of the executioner, or the knife of the aſſaſſin have left him without a brother, the throne of the Sovereign is ſeldom firmly eſtabliſhed.

Such, in the Mahometan empires of Bornou and of Cashna, is the rule of succession to the monarchy; but the Pagan kingdoms adjoining, with obviously less wisdom, permit the several sons of the late Sovereign, attended by their respective partizans, to offer themselves in person, to the choice of the electors, and to be actually present at the decision; an imprudence that often brings with it the interference of other States, and unites the different calamities of foreign and intestine war.

Those of the Royal Children of Bornou who are too young to take share in the reserved part of their deceased father's possessions, are educated in the Palace till the age of maturity arrives; at which time their respective portions of lands and cattle are assigned them.

To

To the four lawful wives of the late Sovereign, a ſeparate houſe, with a ſuitable eſtabliſhment, is granted by the reigning Monarch; and ſuch of his numerous concubines as were not ſlaves, are at liberty to return to their ſeveral friends; and, together with leave to retain their cloaths, and all their ornaments, which are often valuable, have free permiſſion to marry.

In the empire of Bornou, as in all the Mahometan States, the adminiſtration of the provinces is committed to Governors, appointed by the Crown; and the expences of the Sovereign are partly defrayed by his hereditary lands, and partly by taxes levied on the people.

The preſent Sultan, whoſe name is Alli, is a man of an unoſtentatious plain appearance; for he ſeldom wears any other drefs than the common blue fhirt of cotton or of filk, and the filk or muflin turban, which form the ufual drefs of the country. Such, however, is the magnificence of his feraglio, that the ladies who inhabit it are faid to be five hundred in number; and he himfelf is defcribed as the reputed father of three hundred and fifty children, of whom three hundred are males; a difproportion which naturally fuggefts the idea that the mother, preferring to the gratification of natural affection, the joy of feeing herfelf the fuppofed parent of a future candidate for the empire, fometimes exchanges her female child for the male offspring of a ftranger.

Equally fplendid in his ftable, he is faid to have 500 horfes for his own ufe, and for that of the numerous fervants of his houfehold.

In

In many of the neighbouring kingdoms, the Monarch himſelf is the executioner of thoſe criminals on whom his own voice has pronounced ſentence of death: but the Sultan of Bornou, too poliſhed, or too humane, to pollute his hands with the blood of his ſubjects, commits the care of the execution to the Cadi, who directs his ſlaves to ſtrike off the head of the priſoner.

The *Military* Force of the Sultan of Bornou conſiſts in the multitude of his horſemen; for his foot ſoldiers are few in number, and are ſcarcely conſidered as contributing to the ſtrength of the battle.* The ſabre, the lance, the pike,

* The Sultan of Bornoo commands a vaſt army of horſemen, and is a much more powerful Monarch than the Emperor of Morocco.

BEN ALLI.

and the bow, conſtitute their weapons of offence; and a ſhield of hides compoſes their defenſive armour. Fire-arms, though not entirely unknown to them, (for thoſe with which the Merchants of Fezzan occaſionally travel, are ſufficient to give them an idea of their importance and deciſive effect) are neither uſed nor poſſeſſed by the people of Bornou.

When the Sovereign prepares for war, and levies an army for the purpoſe, he is ſaid to have a cuſtom, (the reſult of idle vanity or of politic oſtentation) of directing a date tree to be placed as a threſhold to one of the gates of his capital, and of commanding his horſemen to enter the town one by one, that the parting of the tree in the middle, when worn through by the trampling of the horſes, may enable him to judge of the

ſuffi-

ſufficiency of their numbers, and operate as a ſignal that his levy is complete.

In their *Manners*, the people of Bornou are ſingularly courteous and humane. They will not paſs a ſtranger on the road till they have ſtopped and ſaluted him: the moſt violent of their quarrels are only conteſts of words; and though a part of the buſineſs of their huſbandry is aſſigned to the women, yet, as their employment is confined to that of dropping the ſeed in the furrows, and of removing the weeds with a hoe, it has more of the amuſement of occaſional occupation, than of the harſhneſs of continued labour.

Paſſionately attached to the tumultuous gratifications of play, yet unacquainted with any game but drafts, they often ſit down on the ground, and forming holes to anſwer the purpoſe of ſquares, ſupply the place of men with dates, or the meaner ſubſtitute of ſtones, or of camel's dung. On their ſkill in the management of theſe rude inſtruments of the game, they ſtake their gold duſt, their braſs money, and even their very clothes; and as the bye-ſtanders on theſe occaſions conſtantly obtrude their advice, and ſometimes make the moves for the perſon whoſe ſucceſs they wiſh, their play is uſually accompanied by that conflict of abuſe, and vehemence of ſcolding, which mark and terminate the ſharpeſt of their quarrels.

Such is the amuſement of the lower claſſes of the people: thoſe of a ſuperior rank are devoted to the more difficult

and

and more intereſting game of cheſs, in which they are eminently ſkilled.

In countries that afford without cultivation, or that give in return for ſlight exertions of labour, the principal requiſites of life, few articles of export are likely to be found. Theſe of the Bornou Empire conſiſt of—

Gold Duſt,*	Oſtrich Feathers,
Slaves,	Salt, and
Horſes,	Civet.

By what means the gold duſt, that appears to be a principal article of trade, is procured by the inhabitants, whether from mines in the country, or by purchaſe from other nations, the Shereef has

* At Bornoo I exchanged for gold duſt and oſtrich feathers the merchandize which I brought from Tripoli.

BEN ALLI.

has not explained. But of their mode of obtaining the Slaves, who constitute another extensive branch of their commerce, he gives the following account:

South East of Bornou, at the distance of about twenty days travelling, and separated from it dy several small desarts, is situated an extensive kingdom of the name of Begarmee, the inhabitants of which are rigid Mahometans, and though perfectly black in their complexions, are not of the Negro cast. Beyond this kingdom to the East are several tribes of Negros, idolaters in their religion, savage in their manners, and accustomed, it is said, to feed on human flesh. They are called the Kardee, the Serrowah, the Showva, the Battah, and the Mulgui. These nations the Bergarmeese, who fight on horseback, and are great warriors, annually invade; and when they have

have taken as many prisoners as the opportunity affords, or their purpose may require, they drive the captives, like cattle, to Begarmee. It is said that if any of them, weakened by age, or exhausted by fatigue, happen to linger in their pace, one of the horsemen seizes on the oldest, and cutting off his arm, uses it as a club to drive on the rest.

From Begarmee they are sent to Bornou,* where they are sold at a low price; and thence many of them are conveyed to Fezzan, where they generally emdrace

* The Sultan of Bornoo is continually at war with the rarious idolatrous tribes of Blacks who border on his deminions. Those who are taken prisoners are sold to the Arabs, anb this traffic constitutes the principal commerce of the country.

Slaves are every day brought to him; for the acquisition of this sort of plunder is his constânt occupation.

Ben Alli.

brace the Muſſelman faith, and are afterwards exported by the way of Tripoli to different parts of the Levant.

Such is the mode of obtaining the greateſt part of the ſlaves who are annually ſold in Bornou: but as ſeveral of the provinces of the empire are inhabited by Negros, their inſurrections, real or pretended, afford to the Sovereign an opportunity of increaſing his income by their ſale.

A more politic and more effectual mode of aiding his finances is fruitleſsly offered by the ſalt lakes of the Province of Domboo: for, as the great Empire of Caſhna is entirely deſtitute of ſalt, and none is found in the dominions of the Negros, the ſole poſſeſſion of this article might inſure to the King of Bornou a conſtant and ample revenue of the beſt

kind,

kind, a revenue collected from the ſubjects of Foreign States. But ſuch is the prevalency of antient cuſtom over the obvious ſuggeſtions of policy, that the people of Agadez, a province of the Caſhna Empire, are annually permitted to load their immenſe caravans with the ſalt of Bornou, and to engroſs the profits of this invaluable trade. The ſalt is collected on the ſhores of the ſeveral lakes which produce it, and the only acknowledgement that the Merchants of Agadez give in return for the article, is the trifling price which they pay in braſs and copper (the currency of Bornou) to the neighbouring peaſants.

The civet, which forms another article of the export trade of Bornou, and the greateſt part of which is ſent to the Negro States who inhabit far to the South, is obtained from a ſpecies of wild

cat

cat that is common in the woods of Bornou and of Cashna.

This animal is taken alive in a trap prepared for the purpose, is placed in a cage, and is strongly irritated till a copious perspiration is produced. Its sweat, and especially the moisture that appears upon the tail, is then scraped off, is preserved in a bladder, and constitutes the much valued perfume. After a short interval the operation is renewed, and is repeated, from time to time, till at the end of twelve or fourteen days the animal dies of the fatigue and continual torment. The quantity obtained from one cat is generally about half an ounce.

Of *Manufactures*, none for exportation are furnished by the people of Bornou; but the Shereef remarks that, for their own consumption, they fabricate from the

the iron ore of their country, though with little ſkill, ſuch ſlight tools as their huſbandry requires.*

In return for their exports, they receive the following goods:

Copper

* From the hemp of the country, a coarſe linen is manufactured by the people of Bornoo. Their cotton, which is alſo a native produce, is ſpun to a thread of remarkable fineneſs, and is then converted to callicoes and muſlins of about nine inches in breadth, and of a length which varies from fifteen to twenty yards. Such of theſe cotton manufactures as are enriched with the blue dye of the country, which, from the ſuperiority of the indigo, is preferable to that of the Eaſt Indies, are valued more highly than ſilk; yet their only ſupply of the latter is that which the Merchants of Barbary convey.

They alſo fabricate a ſpecies of carpet, as a covering for their horſes. Tents, from wool and the hair of goats and of camels, are made for the uſe of the army.

The little ſilver they have is converted by their own artiſts into rings.

Ben Alli.

Copper and Braſs, which are brought to them from Tripoli, by the way of Fezzan, and which, as already mentioned, are uſed as the current ſpecies of Bornou;

Imperial Dollars, which are alſo brought to them from Tripoli by the Merchants of Fezzan, and are converted by their own artiſts into rings and bracelets for their women;

Red Woollen Caps, which are worn under the turban;

Check Linens,
Light coarſe Woollen Cloths,
Baize,
Barakans,
Small Turkey Carpets,
Plain Meſurata Carpets.

CHAPTER

CHAPTER VII.

Rout from Mourzouk to Caſhna—Boundaries of the Empire—Its Language, Currency, and Trade.

EQUALLY connected by their commerce with Caſhna and Bornou, the Fezzaners diſpatch to the former as well as to the latter, and always at the ſame ſeaſon, an annual caravan. From Mourzouk, their capital, which they leave at the cloſe of October, they take their courſe to the South South Weſt, and proceed to the Province of Hiatts, the moſt barren, and the worſt inhabited diſtrict of their country.

Five of the fourteen days which are requiſite for this part of their rout, are conſumed in the paſſage of a ſandy deſart, in which their uſual expedient of covering their goat ſkins, both within and without, with a reſinous ſubſtance, prevents but imperfectly the dreaded evaporation of their water,

From the Province of Hiatts they croſs the low mountains of Eyré, which ſeparate the Kingdom of Fezzan from the vaſt Empire of Caſhna; and leaving to their right the ſmall river which flows from theſe hills, and is loſt in the deep ſands of a neighbouring deſart, they enter a wide heath, uninhabited, but not deſtitute of water. The ſixth day conducts them from this extended ſolitude

to

to the long desired refreshments of the Town of Ganatt, where the two next days are devoted to repose.

From thence, by a march of nineteen days, during six of which they are immersed in the heats of a thirsty desart, they pass on to the Town of Assouda, which offers them equal refreshments with Ganatt, and equally suspends their journey.

On leaving Assouda, they travel a delightful country, as fertile as it is numerously peopled; and while the exhilarating sight of Indian corn and of frequent herds of cattle accompanies and chears their passage, the eighth day introduces them to the large and populous City of Agadez, the capital of an extensive province.

Diſtinguiſhed as the moſt commercial of all the towns of Caſhna, and, like Aſſouda and Ganatt, inhabited by Mahometans alone, Agadez naturally attracts the peculiar attention of the Merchants of Fezzan. Many of them proceed no farther; but the greateſt part, committing to their Agents the care of the ſlaves, cotton, and ſenna, which they purchaſe in the courſe of a ten days reſidence, continue their journey to the South.

In this manner, if the camels are compleatly loaded, ſeven and forty days, excluſive of thoſe which are allotted to refreſhment and neceſſary reſt, are employed in travelling from Mourzouk to Agadez.

At the end of three days more, amidſt fields that are enriched with the luxuriant

riant growth of Indian corn, and paſ-tures that are covered with multitudes of cows, and with flocks of ſheep and goats, the Traveller reaches the ſmall Town of Begzam; from which, through a country of herdſmen, whoſe dwellings are in tents of hides, the ſecond day conducts him to the Town of Tegomáh. There, as he ſurveys the ſtony, uninhabited, deſolate hills that form the chearleſs proſpect before him, he caſts a regretful eye on thoſe verdant ſcenes that ſurrounded him the day before. Employed for two days in the paſſage of theſe dreary heights, he deſcends on the third to a deep and ſcorching ſand, from which he emerges at the approach of the fifth evening, and entering a beautiful country, as pleaſingly diverſified with the natural beauties of hills and vales and woods, as with the rich rewards of the huſbandman's and the ſhepherd's toil, he arrives in ſeven days more at the City of Caſhna, the capital of the empire of which it bears the name, and the uſual reſidence of its powerful Sultan.

The country to which the Geographers of Europe have given the name of Nigritia, is called by the Arabs Soudán, and by the natives Aafnou, two words of ſimilar import, that, like the European appellation, expreſs the land of the Blacks, and like that too, are applied to a part only of the region to which their meaning ſo obviouſly belongs.—Yet, even in this limited ſenſe, the word Soudán is often variouſly employed; for while ſome of the Africans reſtrict it to the Empire of Caſhna, which is ſituated at the North of the Niger, others extend it, with indefinite comprehenſion, to the Negro States on the South

South of the river, and applying it as a means of expreſſing the extended rule and tranſcendant power of the Emperor of Caſhna; call him, with extravagant compliment; the Sultan of all Soudán.

His real ſovereignty is bounded, on the North, by the mountains of Eyré, and by one of thoſe diſtricts of the great Zahara, that furniſh no means of uſeful property or available dominion; on the South, by the Niger; and on the Eaſt, by the Kingdom of Zamphara and the Empire of Bornou, Its weſtern limit is not deſcribed by the Shereef; nor is any thing ſaid of the Capital, except that it is ſituated to the North of the Niger, at the diſtance of five days journey, and that its buildings reſemble thoſe of Bornou.

The obſervations which introduced the account of Bornou, have already announced the remarkable ſimilarity, as well with reſpect to climate, ſoil, and natural productions, as with regard to the colour, genius, religion, and political inſtitutions of the people, that prevails between that powerful State and its ſiſter Kingdom of Caſhna.

The *rains*, indeed, are leſs violent than thoſe of Bornou. It excluſively furniſhes the biſhnah, a ſpecies of Indian corn that differs from the gamphúly, in the blended colours of red and white which diſtinguiſh its grain. Its monkeys and parrots (animals but ſeldom ſeen in Bornou) are numerous, and of various ſpecies. The meridian of its capital is conſidered as a weſtern limit, in that parallel of latitude, to the vegetation of of grapes and the breed of camels; for between Caſhna and the Atlantic few camels are bred, and no grapes will grow. The manners of the common people are leſs courteous in Caſhna than in Bornou, and their games are leſs expreſſive of reflection; for their favourite play conſiſts in toſſing up four ſmall ſticks, and counting thoſe that croſs each other, as ſo many points of the number that conſtitutes the game. But the circumſtances of chief diſcrimination between the empires are, thoſe of language, currency, and certain articles of commerce.

Of the difference between the *Languages* of Bornou and of Caſhna, the following ſpecimen is given by the Shereef.

In the Language of Bornou		In the Language of Caſhna	
1	*is expreſſed by* Lakkah	1	*is expreſſed by* Deiyrh
2	.... Endee	2	.... Beeyou
3	.... Nieſkoo	3	.... Okoo
4	.... Dokoo	4	.... Foodoo
5	.... Okoo	5	.... Beàt
6	.... Araſkoo	6	.... Shedah
7	.... Nuſkoo	7	.... Bookai
8	.... Tallóre	8	.... Tâkoos
9	.... L'ilkar	9	.... Tarràh
10	.... Meikoo	10	.... Goumah
11	.... Meiko Lakkah	11	.... Goumah ſha Deiyah
12	.... Meiko Endee	12	.... Goumah ſha Beeyou
13	.... Meikoo Nieſkoo	13	.... Goumah ſha Okoo
14	.... Meikoo Dekoo	14	.... Goumah ſha Foodoo

The *Currency* of Caſhna, like that of the Negro States to the South of the Niger, is compoſed of thoſe ſmall ſhells that

that are known to Europeans and to the Blacks themſelves by the name of Cowries, and to the Arabs by the appellation of Hueddah.—Cardie, which is another term for this ſpecies of Negro money, and the ſpecific meaning of which the Shereef has neglected to explain, is ſaid to be given to it by the idolatrous tribes alone; a circumſtance that ſeems to indicate ſuperſtitious attachment.——Of theſe ſhells, 2,500 are eſtimated in Caſhna as equal in value to a mitkal of Fezzan, which is worth about 675 piaſtres of Tripoli, or ten ſhillings and three half-pence ſterling.

Among the few circumſtances which characterize the *Trade* of Caſhna, as diſtinguiſhed from that of Bornou, the moſt remarkable is, that the Merchants of the *former* kingdom are the ſole carriers, to other nations, of a ſcarce and moſt valuable

able commodity, which is only to be obtained from the inhabitants of the *latter*. For though the ſalt of Bornou ſupplies the conſumption of Caſhna, and of the Negro Kingdoms to the South, yet its owners have abandoned to the commercial activity of the Merchants of Agadez, the whole of that profitable trade.

The lakes, on the dreary ſhores of which this article of African luxury is found, are ſeparated from Agadez by a march of five and forty days, and are encompaſſed on all ſides by the ſands of the vaſt Deſart of Bilma, where the ardent heat of a flaming ſky is returned with double fierceneſs by the ſurface of the burning ſoil. A thouſand camels, bred and maintained for the purpoſe, are ſaid to compoſe the caravan which annually explores, in the ſavage wilderneſs, the long line of this adventurous journey.

journey. Perilous, however, and full of hardſhips as their labour is, the Merchants find an ample recompence in the profits of their commerce; for while the wretched villagers who inhabit the neighbourhood of the lakes, and collect the ſalt that congeals upon the ſhores, are contented to receive, or obliged to accept a ſcanty price, the value that the Merchants obtain in the various markets of Caſhna, of Tombuctou, and of the countries to the South of the Niger, is ſuited to the high eſtimation in which the article is held.

Attentive in this manner to the means of profiting by the produce of a neighbouring country, the people of Agadez are equally anxious to avail themſelves of the commodities that are furniſhed by their own; for knowing the ſuperior quality of the ſenna which grows upon their

their mountains, they demand and receive from the Merchants of Fezzan a proportionable price.—

The ſenna of Agadez is valued in Tripoli at fourteen or fifteen mahaboobs, or from 4l. 4s. to 4l. 10s. per hundred weight; while that of Tibeſti is worth no more than from nine to ten mahaboobs, or from 2l. 14s. to 3l. ſterling. From Tripoli the ſenna is exported to Turkey, Leghorn, and Marſeilles.

Of the other articles of ſale which the extenſive Empire of Caſhna affords, the principal are—

Gold Duſt—the value of which appears to be eſtimated at a higher rate in Caſhna than in Fezzan; for in the former the worth of an ounce of 640 grains (which

(which is the weight of an ounce in Fezzan, Caſhna, and probably in all the States between that kingdom and the Niger) is ſaid to be nine mitkals, or 4l. 10s. ſterling; whereas an ounce of the ſame weight is worth in Fezzan but 4l. In Caſhna the value of an Engliſh ounce of 480 grains is conſequently 3l. 7s. 6d. whereas in Fezzan it is only 3l.

Slaves.—In what manner theſe are obtained, does not diſtinctly appear; but the value of a male ſlave is ſaid to be from 15 to 20,000 cowries, or from 3l. to 4l. ſterling:

That of a female ſlave is deſcribed as being two-thirds of the former, or from 10,000 to 13,334 cowries, which in Engliſh money would be from 2l. to 2l. 13s. 4d.

Cotton

Cotton Cloths—which are the general manufacture of Caſhna, of Bornou, and of the Negro States to the South of the Niger:

Goat-ſkins—of the red and of the yellow dyes:

Ox and Buffalo Hides:

Civet—the mode of obtaining which, as well as the principal markets for its ſale, were deſcribed in the account which has been given of the trade of Bornou.

In return for theſe articles the inhabitants of Caſhna receive—

Cowries—a ſea ſhell which is brought from the coaſt, and conſtitutes the common ſpecie of the empire:

Horſes

Horſes and Mares—which are purchaſed from the Merchants of Fezzan; but whether bred in that country, or procured from the Arabs, or from the people of Bornou, is not mentioned by the Shereef:

Red Woollen Caps,
Check Linens,*
Light coarſe Woollen Cloths,
Baize,
Barakans or Alhaiks,
Small Turkey Carpets,
Plain Meſurata Carpets,
Silk, wrought and unwrought,
Tiſſues and Brocades,
Sabre Blades,

S Dutch

* The dreſs of the people of Fullan (a country to the Weſt of Caſhna) reſembles the cloth of which the plaids of the Scotch Highlanders are made.

BEN ALLI.

Dutch Knives,

Sciſſars,

Coral,

Beads,

Small Looking-Glaſſes,

Tickera—a paſte which is prepared in Fezzan from dates and the meal of Indian corn, and which, whenever they travel, is in great requeſt among the people of Fezzan:

Gooroo Nuts——which are brought from the Negro States on the South of the Niger, and which are principally valued for the pleaſant bitter that they communicate to any liquid in which they are infuſed.

CHAPTER

CHAPTER VIII.

Countries South of the Niger.

THE account which the Shereef has given of ſuch of the kingdoms to the South of the Niger as he himſelf has viſited, is too deficient in geographical information, to furniſh a clear and determinate idea of this part of his travels: and though the names of the principal States in whoſe capitals he traded, or through whoſe dominions he paſſed, may be uſed with advantage as the means of future enquiry, and are therefore inſerted in the map which accompanies

 this

this Narrative; yet the places aſſigned them muſt be conſidered as in ſome degree conjectural. That the line of his journey was towards the Gold Coaſt, there is, however, the ſtrongeſt reaſon to believe; and the following brief account of his remarks may lead to concluſions which are neither unintereſting nor unimportant.

From that part of the Niger which forms the ſouthern limit of the great Empire of Caſhna, to the Kingdom of Tonouwah, which borders on the coaſt of the Chriſtians, and of which the Town of Aſſenté is ſaid to be the capital, a ſucceſſion of hills, among which are mountains of a ſtupendous height, diverſifies or conſtitutes the general face of the country. Moſt of the lands are deſcribed as already cleared, but ſome particular diſtricts are ſtill incumbered with woods of of a vaſt extent; and though for the moſt part the highlands are paſtured by innumerable flocks of ſheep, and by herds of cows and goats, and the vales exhibit the captivating view of ſucceſſive villages, encompaſſed with corn and rice, and fruits of various kinds, yet there *are* places of native ſterility and eternal barrenneſs.

The combined occupations of the ſhepherd and of the huſbandman compoſe the general employment of the people; while the cotton cloth, and the goatſkins of the red and of the yellow dyes, that are offered in ſeveral of the towns for ſale, announce the rudiments of future manufactures, and perhaps of an extenſive commerce.

Exempted by the nature of their climate from many of thoſe hardſhips from

 which,

which, in other countries, dreſs is the principal protection, a large proportion of the inhabitants wear only the covering that decency requires. But moſt of the Mahometans, as the mark of a religion which they are proud to profeſs, adorn their heads with the folds of the turban, and alſo adopt, at leaſt in ſome of the States, the cotton ſhirt, which is ſo much worn in the empires of Caſhna and Bornou.

Tents, which are formed of the hides of cows or of buffaloes, and which are peculiarly ſuited to the ſhepherd life, are the only dwellings of multitudes of the Negros; while the huts, which others erect with the branches of trees, are of a conſtruction almoſt equally ſimple.

Several of the towns are deſcribed as ſurrounded

ſurrounded by walls; and bows and arrows are mentioned as the common inſtruments of war.

In the deſcription of their Government, a few inſtances of ſmall Republics are given; but moſt of their States are monarchical; and of theſe, the inhabitants of the Mahometan Kingdom of Degombah are diſtinguiſhed by the cuſtom of taming the Elephant, and by that of ſelling for ſlaves the priſoners they take from ſuch of the bordering nations as motives of religion or of avarice prompt them to invade.

Such, however, is the mildneſs of the Negro character, that even the aſperities of religious diſagreement appear to have no effect on their general conduct; for there is reaſon to believe, from the Shereef's account, that the Muſſelman and the Pagan are indiſcriminately mixed; that their cattle feed upon the ſame mountain; and that the approach of evening ſends them in peace to the ſame village: and though the nations who are attacked by the people of Degombah puniſh with death, as guilty of atrocious injuſtice, ſuch of the invaders as the chance of war throws into their hands, yet thoſe of the Mahometans who viſit them for the purpoſes of trade, are received with protection and reſpect.

To the Merchants of Fezzan, who travel to the ſouthern States of the Negros, the purchaſe of gold, which the dominions of ſeveral, and eſpecially of thoſe of Degombah, abundantly afford, is always the firſt object of commercial acquiſition. The other articles which they obtain, conſiſt of

Slaves,

Cotton

Cotton Cloth,

Goat-Skins, of a beautiful dye,

Hides of Buffaloes and Cows, and

A ſpecies of Nut—which is much valued in the kingdoms to the North of the Niger, and which is called Gooroo. It grows on a large and broad leafed tree that bears a pod of about eighteen inches in length, in which are incloſed a number of nuts that varies from ſeven to nine. Their colour is a yellowiſh green; their ſize is that of a cheſnut, which they alſo reſemble in being covered by a huſk of a ſimilar thickneſs; and their taſte, which is deſcribed as a pleaſant bitter, is ſo grateful to thoſe who are accuſtomed to its uſe, and ſo important as a corrective to the unpalatable or unwholeſome waters of Fezzan, and of the other kingdoms that border on the vaſt Zahara, as to be deemed of importance to the happineſs of life.

No

No commercial value appears to be annexed to the fleeces which the numerous flocks of the Negro kingdoms afford; for the cotton manufacture, which, the Shereef ſays, is eſtabliſhed among the tribes to the South of the Niger, ſeems to be the only ſpecies of weaving that is known among them. Perhaps the dark colour of the fleece, as diſqualifying it for the dye, may be one reaſon, and its coarſe and hairy nature may be another, of the little eſteem in which it appears to be held.

In *return* for the articles which they ſell to their foreign viſitors, the Negros receive—

Salt, from the Merchants of Agadez,
Dutch Knives,
Sabre Blades,
Carpets,
Coral,

Beads,

Beads,

Looking-Glaſſes,

Civet,

Imperial Dollars and Braſs--from both of which the Negro artiſts manufacture rings and bracelets for their women.

Fire Arms are unknown to ſuch of the nations on the South of the Niger as the Shereef has viſited; and the reaſon which he aſſigns for it is, that the Kings in the neigbourhood of the coaſt, perſuaded that if theſe powerful inſtruments of war ſhould reach the poſſeſſion of the populous inland States, their own independence would be loſt, have ſtrictly prohibited, and by the wiſdom of their meaſures, have effectually prevented this dangerous merchandize from paſſing beyond the limit of their dominions.

CHAPTER

CHAPTER IX.

General View of the Trade from Fezzan to Tripoli, Bornou, Cashna, and the Countries on the South of the Niger.

IN the general defcription of Fezzan, an account was given of the various articles of native produce which fupply the wants, or contribute to the trade of its people; but of their *Foreign* Commerce, for which, like the Dutch in Europe, they are eminently diftinguifhed, the detail was purpofely deferred: for till a previous account of the countries

to

to which that commerce is eſtabliſhed had been exhibited, no adequate conception of its nature or extent could be eaſily conveyed.

At the latter end of October, when the ardent heat of the Summer months is ſucceeded by the pleaſant mildneſs and ſettled ſerenity of Autumn, the ſeveral caravans that are reſpectively deſtined for Tripoli and Bornou and Caſhna, and the Negro Nations beyond the Niger, take their departure from Mourzouk, the capital of Fezzan. The parties which compoſe them are generally ſmall; for unleſs information has been received that the road is infeſted with robbers, ten or a dozen Merchants attended by twice as many camels, and by the neceſſary ſervants, conſtitute the uſual ſtrength of the caravan. But if an attack is apprehended, an aſſociation of forty

forty or fifty men, with mufkets for their defence, is formed; and as none of the Africans to the South of Fezzan (the people of Agadez and the nations on the coaft excepted) have yet poffeffed themfelves of fire-arms, the collective force of fuch a number is fufficient to infure their fafety.

Their ftore of provifions ufually confifts of dates; of meal prepared from barley or from Indian corn, and previoufly deprived of all its moifture in an oven temperately heated; and of mutton, which is cured for the purpofe, by the treble procefs of being falted and dried in the fun, and afterwards boiled in oil or fat; a procefs which gives it, even in that climate, a lafting prefervation.

In all the principal towns to which they

they trade, the Merchants of Fezzan have Factors, or confidential Friends, to whose care, till their return, or till their instructions as to the market shall arrive, they consign such Negros as they purchase, perfectly assured that the slaves will be forwarded by the Agents according to the orders they receive; but their gold dust, as being more easily conveyed, and less dependant for its value on the choice of the market, is seldom entrusted to the Factor.

The caravans which proceed to *Tripoli* are freighted partly with trona, the produce of their native land, and partly with senna and gold dust and slaves, the produce of the southern countries with which they trade: and in return they bring back the cutlery and woollens (particularly red woollen caps) and silks, wrought and unwrought, together with the

the Imperial dollars, the copper and the braſs, which are requiſite for the conſumption of thoſe countries or for their own.

The caravans which travel to *Bornou* are loaded with the following goods:

Braſs and copper—for the currency of Bornou. The caravan which Mr. LUCAS accompanied from Tripoli to Meſurata, had brought ten camel loads or forty hundred weight of theſe metals for the Bornou market: their value in Bornou is about four ſhillings ſterling for each pound weight.

Imperial Dollars—which are called in Arabic Real Abotacia, and the value of which, in compariſon with the dollars of Spain, is, at Tripoli, as 365 piaſtres to 340, or nearly as 16 to 15:

Red Woollen Caps,
Check Linens,
Light coarſe Cloth,
Baize,
Barakans or Alhaiks,
Small Carpets of Turkey,
Small plain Carpets of Meſurata,
Silk, wrought and unwrought,
Tiſſues and Broçades—for the Royal Family and other perſons of rank,
Sabre Blades,
Dutch Knives,
Sciſſars,
Coral,
Beads,
Small Looking-Glaſſes,
Gooroo Nuts.

Of the native produce of Fezzan the only article which is brought as merchandize to Bornou is a preparation of pounded dates, and of the meal of Indian corn,

corn, highly dried in an oven. It is called Tickera, and is valued, eſpecially by Travellers, as a portable and highly ſalubrious food.

In *return* for the goods which they bring to *Bornou* the Merchants *take back* with them,

Slaves,

Gold Duſt,

Civet—for the markets on the South of the Niger.

The *exports* from Fezzan to *Caſhna* and its dependent States, conſiſt of the following articles;

Cowries—a ſea ſhell (in Arabic called Hueddah) which conſtitutes the circulating ſpecie of this empire, and of the

Negro kingdoms, and which the Merchants procure from the Southern nations who border on the coaſt; 17,062 are conſidered in Caſhna as equivalent to an Engliſh ounce (480 grains) of gold;

Braſs—from which the Smiths of the country manufacture rings and bracelets for their women;

Horſes,
Red Woollen Caps,
Check Linens,
Light coarſe Cloths,
Baize,
Barakans, or Alhaiks,
Small Turkey Carpets,
Plain Meſurata Carpets,
Silk, wrought and unwrought,
Tiſſues and Brocades,
Sabre Blades,

Dutch

Dutch Knives,

Sciſſars,

Coral,

Beads,

Small Looking-Glaſſes,

Tickera—a preparation of pounded dates, and the meal of Indian corn, which is manufactured in Fezzan:

Gooroo Nuts—which are brought from the Negro Countries on the South of the Niger, and the uſe of which has been mentioned before.

The articles received in *return*, are—

Gold Duſt—of which an Engliſh ounce (or 480 grains) appears to be valued at 3l. 8s. 3d. though in Fezzan it ſeems to be worth no more than 3l. The Fezzaners, in all probability, make themſelves amends by the price which they charge upon their goods:

Slaves—A male ſlave is worth in Caſhna, from 3l. 10s. to 5l.—a female ſlave is worth two-thirds of the amount, or from 2l. 6s. 8d. to 3l. 6s. 8d.

Cotton Cloths—of various colours, principally blue and white, of which in the Empire of Caſhna, and in the Negro States to the South of the Niger, great quantities are made:

Goat Skins—dyed red or yellow:
Ox and Buffalo Hides—for tents:

Senna from Agadez—a province of the Caſhna Empire:—the Agadez ſenna is worth at Tripoli, from fourteen to fifteen mahaboobs (4l. 4s. to 4l. 10s. ſterling)

ling) per hundred weight: that which the Fezzaners obtain at Tibefti is only worth per hundred weight, from nine to ten mahaboobs, or from 2l. 14s. to 3l. fterling.

Civet.

To fuch of the various nations inhabiting *the Country on the South of the Niger* as they are accuftomed to vifit, the Merchants of Fezzan convey the following articles:

Sabre blades,
Dutch Knives,
Carpets,
Coral,
Beads,
Looking-Glaffes,
Brafs,

Imperial Dollars,
Civet.

In *return* the Merchants receive—

Gold Duſt,
Slaves,
Cotton Cloths—of various colours,
Goat Skins—red and yellow,
Ox and Buffalo Hides,
Gooroo Nuts—for ſale in Caſhna, Bornou, and Fezzan, where they are purchaſed at the rate of 12s. for one hundred pods:
Cowries—for ſale in Caſhna.

Ivory, though very common in the country to the South of the Niger, is not conſidered by the Merchants of Fezzan as an article of profitable tranſport, the demand for it on the Coaſt being ſuch as induces them to ſell to the Negros who traffic there, the teeth which in the

the courſe of their journey, they often find in the woods.*

Such

* The goods *imported* by the Merchants of Fezzan, conſiſt of

Slaves—in which they have a great trade,

Gold Duſt,

Ivory,

Oſtrich Feathers,

Senna—which is brought from the neighbouring countries,

European goods of various ſpecies, from Tripoli.

The goods *exported* from Fezzan, conſiſt of the following European articles:

Woollen Cloths,

Linens of different ſorts,

Gun Barrels,

Gun and Piſtol Locks,

Small Shot,

Ball,

Iron Bars,

Tin,

Copper,

Such are the principal branches of the extenſive commerce of the Merchants of Fezzan; from a view of which it appears, that, vaſt as their concerns are, they have little communication with any of the States that are ſituated to the Weſt of the Empire of Caſhna; a circumſtance which the Shereef aſcribes to the want

---

Copper,
Braſs,
Braſs Diſhes,
Nails,
Spices,
Muſk,
Benzoin,
Dying Wood,
Allum,
Tartar for Dying,
Green Vitriol,
Verdigreaſe,
Brimſtone,
Looking-Glaſſes.

BEN ALLI.

want of a proper conveyance for their goods; for the country on the Weſt of Caſhna furniſhes but few camels, and even horſes and mules are ſingularly ſcarce and dear.

CHAPTER X.

Rout from Mourzouk to Grand Cairo, according to Hadgee Abdalah Benmileitan, the present Governor of Mesurata.

PLACED in a situation which affords an easy intercourse with the Mediterranean, and therefore with the States of Europe, on the one hand, and on the other with the extensive Empires of Bornou and Cashna, the dominions of of Tombuctou, and the various nations of Negros to the South of the Niger, the Merchants of Fezzan are happily possessed

poſſeſſed of the farther advantage of communicating by a ſafe and comparatively commodious paſſage with the Cities of Grand Cairo and of Mecca. A pilgrimage to the latter, the object, from time immemorial, of veneration in Arabia, is preſcribed to every Muſſelman; and though the greateſt part of the believers in Mahomet, deterred by diſtance, or reſtrained by the avocations of buſineſs and the feelings of domeſtic attachment, content themſelves with imperfect reſolutions of performing at ſome future period this arduous journey; yet there are perſons, even from the innermoſt receſſes of Africa, who think, that a poſitive injunction of their faith is too ſolemn for excuſes, and too momentous for delay. Prompted by this urgent conſideration, or allured by the honourable diſtinction which attends upon the title of Hadgee, the envied, appellation

pellation of thoſe who have viſited the ſacred Temple, a number of the faithful from the Empires of Bornou and Caſhna, from the extenſive kingdom of Caſſaba, and from ſeveral of the Negro States, reſort to Fezzan, and proceed from thence, with the caravan, which in the Autumn of every ſecond or third year takes its departure for Mecca. The caravan, which ſeldom conſiſts of leſs than one hundred, or of more than three hundred Travellers, aſſembles at Mourzouk, and begins its journey in the laſt week of October, or in the firſt of the ſucceeding month.

Temiſſa, a town in the dominions of Fezzan, and ſituated to the Eaſt North Eaſt of Mourzouk, receives them at the cloſe of the ſeventh day; and in two days more, of eaſy travelling, they arrive at a lofty mountain, rocky, uninhabited,

bited, and barren, of the name of Xanibba. Having recruited their goat-ſkin bags from the only well which theſe ſullen heights afford, they deſcend to a vaſt and dreary deſart, whoſe hilly ſurface, for four ſucceſſive days, preſents nothing to the eye but one continued extent of black and naked rock; to which, for three days more, the equally barren view of a ſoft and ſandy ſtone ſucceeds. Through all this wide expanſe of varied nakedneſs no trace of animal or vegetable life, not even the deſart thorn, is ſeen. On the eighth day, the vaſt mountain of Ziltan, the rugged ſides of which are marked with ſcanty ſpots of bruſhwood, and are enriched with ſtores of water, increaſes the labour of the journey. Four days are devoted to the toils of this ſtupendous paſſage; four others are employed in croſſing the ſultry plain that ſtretches its barren ſands

from

from the foot of the mountain to the verdant heights of Sibbeel, where the wells of water and the chearing view of multitudes of antelopes ſuſpend their fatigues, and anticipate the refreſhments that await them on the next evening; for the cloſe of the following day conducts them to the town of Augéla.

From that place, which is ſubject to Tripoli, and is famed for the abundance and excellent flavour of its dates, they proceed in one day to the little village of Gui Xarrah: another brings them to the long aſcent of the broad mountain of Gerdóbah, from whoſe inflexible barrenneſs the Traveller, in the courſe of a five days paſſage, can only collect a ſcanty ſupply of unpalatable water. Deſcending from theſe mournful highlands, he enters the narrow plain of Gegabib, ſandy and uninhabited, yet fertile in dates, which the people of Duna (a town dependant on Tripoli, and ſituated on the Coaſt at the diſtance of eight days journey from Gegabib) annually gather.

From this ſcene of gladſome contraſt to the inveterate rocks of Gerdóbah, a three days march conducts the caravan to another deſolate mountain of the name of Buſelema, that furniſhes only water; and in three days more they enter the dominions of the independent Republic of See-wah.

Governed by a Council of ſix or eight Elders, whoſe laſting diſſentions divide the opinions and diſtract the allegiance of the people, this unfortunate State is conſtantly involved in the miſeries of inteſtine war. Its chief produce is the date

date tree; for the lands, though not destitute of water, furnish but little corn.

From See-wah, the capital, the caravan proceeds in a single day to the miserable village of Umseguér, which is one of the dependencies of the State, and is situated at the foot of the mountainous Desart of Le Mágra, where, in the long course of a seven days passage, the Traveller is scarcely sensible that a few spots of thin and meagre brushwood slightly interrupt the vast expanse of sterility, and diminish the amplitude of desolation. The eighth day terminates with his arrival at the hill of Huaddy L'Ottrón, which is distinguished by a small convent, of three Christian Monks, who reside there under the protection of Cairo, and to whose hospitable entertainment the Traveller is largely indebted. Buildings surrounded with high walls, and erected in the neighbourhood of the convent, are opened for his reception; and for three successive days, if he chuses to be their guest so long, his wants, as far as their means extend, are chearfully and liberally supplied.

Their garden, in which is a well of excellent and never-failing water, affords an ample store of vegetables of various kinds: the maintenance of a few sheep is furnished by an adjoining pasture; and they raise, without difficulty, a numerous breed of fowls. All other articles, except their bread, which they manufacture themselves, they receive from Cairo.

Respected by the Arabs, who revere their hospitality more than they hate their religion, these venerable men are apparently secure.——Yet as too much con-

confidence might invite the meaneſt plunderers to invade their peaceful dwelling, they have cautiouſly guarded their convent by a ſeparate and lofty incloſure from an opening in which a ladder of ropes furniſhes the means of deſcent.

Leaving this hoſpitable hill with ſuch refreſhments as the generous Fathers could ſupply, the caravan continues its courſe, and on the fifth day arrives at the City of Cairo, from whence, at the uſual ſeaſon, it proceeds by the cuſtomary rout to Mecca.

CHAPTER XI.

Conclusions suggested by the preceding Narrative.

FROM the perusal of the preceding account, the Society are enabled to judge of the credit which it deserves, and of the value of the information which it offers. If the evidence of its truth should be thought insufficient, they will keep their minds in suspence till Narratives of more confirmed authenticity shall disprove or establish the relation. But if they should think that the Shereef's account of Bornou and of the Niger is too strongly supported by the corresponding description of Ben Alli, to leave a suspicion of its falſehood; and that the fidelity of this part of his Narrative warrants the belief of whatever elſe is equally deſcribed on the authority of his perſonal knowledge, then they will feel that concluſions of an important and intereſting nature reſult from the various, though imperfect intelligence which he has furniſhed.

The preſent ſtate of the Empire of Bornou, compared with its condition when Leo Africanus, who wrote his account in the year 1526, was its viſitor, exhibits an intereſting proof of the advancement of the Mahometan Faith, and of the progreſs of imperfect civilization.* A ſavage nakedneſs, or the rude covering

* *Leo's Hiſt. of Africa, book* vii. *p.* 293, 294. *Eng. Edition.*

" OF THE KINGDOM OF BORNO.

" The inhabitants in Summer go all naked, except at " their waiſts, which they cover with a piece of leather: " but covering which the ſkins of beaſts afford, are now relinquiſhed, for the decency

" but all the Winter they are clad in ſkins, and have beds " of ſkins alſo. They embrace no religion at all, being " neither Chriſtians, Mahometans, nor Jews, nor of any " other profeſſion, but living after a brutiſh manner, and " having wives and children in common: and (as I un- " derſtood of a certain Merchant that abode a long time " among them) they have no proper names at all, but " every one is nick-named according to his length, his " fatneſs, or ſome other qualitie. They have a moſt " puiſſant Prince, being lineally deſcended from the Ly- " bian people called Bardoa: horſemen he hath in a con- " tinual readineſs, to the number of 3000, and an huge " number of footmen, for all his ſubjects are ſo ſervice- " able and obedient unto him, that whenſoever he com- " mandeth them, they will arme themſelves and follow " him whither he pleaſeth to conduct them. They paye " unto him none other tribute but the tithes of all their " corne: neither hath this King any revenues to maintain " his Eſtate, but only ſuch ſpoils as he getteth from his " next enemies by often invaſions and aſſaults."

—" Yet the King ſeemeth to be marvellous rich; for " his

cency and convenience of a drefs of cotton manufacture. Tempered by the courtefy of commerce, and the conciliating interchange of important benefits, the antient barbarifm of the people is foftened to habits of kindnefs; and, in the minds of the greateft part, the abfurd fuperftitions of Paganifm have given place to the natural and fublime idea of the Unity of God.

From the account of the nations to the *South of the Niger*, it appears, that, among

" his fpurs, his bridles, platters, difhes, pots, and other " veffels wherein his meat and drink are brought to the " table, are all of pure gold: yea, and the chains of " his dogs and hounds are of gold alfo. Howbeit this " King is extremely covetous, for he had much rather pay " his debts in flaves than in gold. In this kingdome are " great multitudes of Negros and of other people, the " names of whom (bicaufe I tarried here but one month) " I could not well note."

among the Mahometan Blacks, the ſtern-neſs of the Muſſelman is ſoftened by the mildneſs of the Negro; and that if the zeal of the invader be not inflamed by the value of the captive, the propagation of the faith is ſeldom conſidered as a reaſon or pretext for war.

Perhaps the attention of the *Philoſopher* may be engaged by the uſe to which the ſmall ſhells of the Maldive Iſlands are equally applied by the inhabitants of Caſhna, and by the natives of Bengal. Sameneſs of opinion, or reſemblance of conduct, when founded in natural feeling, or a ſimilar ſtate of ſociety, are ſeen without wonder in nations unconnected and remote; but that a cuſtom ſo arbitrary and artificial as that of employing Cowries as a ſubſtitute for coin, a cuſtom which inſtinct could not have produced, and chance could ſcarcely have

occaſi-

occaſioned, ſhould equally prevail among the Negros of Africa and the natives of Bengal, may juſtly be deemed a curious and intereſting phenomenon.

To the Britiſh *Traveller*, a deſire of exchanging the uſual excurſion from Calais to Naples, for a Tour more extended and important, and of paſſing from ſcenes with which all are acquainted, to reſearches in which every object is new, and each ſtep is diſcovery, may recommend the Kingdom of Fezzan. If Antiquities be his favourite purſuit, the ruins which ſhadow the cottages of Jermah and of Temiſſa, promiſe an ample gratification: or if the ſtudy of Nature be his wiſh, the expanſive ſcenes and numerous productions of that uninveſtigated ſoil may equally promote his entertainment and his knowledge. But if a ſpirit of more adventurous reſearch ſhould

ſhould induce him to travel with the Merchants of Fezzan; diſcoveries of ſuperior value may diſtinguiſh and reward his toil. The powerful Empires of Bornou and Caſhna will be open to his inveſtigation; the luxurious City of Tombuctou, whoſe opulence and ſevere police attract the Merchants of the moſt diſtant States of Africa, will unfold to him the cauſes of her vaſt proſperity; the myſterious Niger will diſcloſe her unknown original and doubtful termination; and countries unveiled to antient or modern reſearch will become familiar to his view. Or ſhould he be willing to join the Cairo Caravan, the diſcovery of the antient ſcite, and of whatever elſe may remain of the Temple of Jupiter Ammon, may perhaps be attempted with ſucceſs: for the ſame cauſes which gave birth to the ſprings, and, by their means, to the luxuriant vege-

vegetation of the antient domains of the Temple, muſt ſtill continue to diſtinguiſh the fortunate ſoil: and there ſeems no reaſon to doubt that the hoſpitable convent of Huaddy L'Ottrón, or the neighbouring Republic of See-wah, will not only furniſh a place of convenient departure and of eaſy retreat, but will alſo ſupply the requiſite information, and the neceſſary aid. At this time, an Interpreter, who is equally acquainted with Arabic and Engliſh, may be found in London; and, at all times, the Cities of Tripoli, of Tunis, and Ceuta, afford a number of Mahometans who are almoſt as converſant with the Italian or the Spaniſh, as with the Arabic, their native language.

But of all the advantages which a better acquaintance with the Inland Regions of Africa may lead, the firſt in im-

portance

portance is, the extenſion of the Commerce, and the encouragement of the Manufactures of Britain. That fire-arms are in requeſt with all nations, civilized and ſavage, who have the means of obtaining them, experience uniformly proves; and we now learn, that to the jealouſy which the Princes on the Southern Coaſt entertain of the powerful Interior States, and to the total neglect of all opportunities of opening, from more favourable quarters, an immediate communication with the inland country, it is owing that the ſale of one of the moſt profitable manufactures of Great Britain is ſtill in a great meaſure confined to the ſcanty Tribes which inhabit the ſhore of the Atlantic. From the ſame information we are alſo led to conclude, that the anxious policy which prohibits the conveyance of fire-arms to the Inland Tribes, dictates, as a neceſſary caution,

tion, the ſevereſt reſtraints on the tranſit of other merchandize. But if, on the ſyſtem of the Moors; the effect of which has been tried too long for its wiſdom to be diſputed, aſſociations of Engliſhmen ſhould form caravans, and take their departure from the higheſt navigable reaches of the Gambia; or from the ſettlement which is lately eſtabliſhed at Sierra Leona; there is reaſon to believe, that countries new to the fabrics of England, and probably inhabited by more than a hundred millions of people, may be gradually opened to her trade. On this ſyſtem, much greater would be their profits than thoſe which the Merchants of Fezzan receive; for they would reach, by a journey of 700 miles from their veſſels, the ſame markets to which a land-carriage of 3000 conveys from the Mediterranean the goods of the Fezzaners; and they would alſo poſſeſs the

farther

farther advantage of obtaining at prime coſt, the ſame articles for which the Merchants of Fezzan are ſubjected to the complicated diſadvantages of a high price, of an inferior quality, and of the varying exactions that the deſpotic Governments of Barbary impoſe. Now if it be conſidered, that notwithſtanding the vaſt expence of land-carriage, and of an exorbitant price on the purchaſe of the articles which they ſell, the Traders of Fezzan are ſtill enabled to collect a profit that upholds and encourages their commerce, it muſt be evident, that the gain which the Merchants of England would derive from a ſimilar traffic, conducted as is here propoſed, would be ſuch as few commercial adventures have ever been found to yield. That no difficulties will attend the execution of the Plan, the general hiſtory of new undertakings forbids us to believe; but as far as the *climate* and *religion* of the Negros are in queſtion, there ſeems to be little diſcouragement; for the long deſcent of the rivers is a proof that the elevation of the inland country is raiſed above the level of the coaſt, and conſequently that the climate is much more temperate, and probably more ſalubrious: and while the Narrative of the Shereef announces that the Merchant is conſidered by the Negro as the general friend of Mankind, common experience convincingly ſhews that, in the judgment of the Pagan, the Creſcent and the Croſs are objects of equal indifference; and that the comparative welcome of the Muſſelman and of the Chriſtian depends on their Merchandize rather than their Creed.

As little diſcouragement is ſuggeſted by a view of the purchaſeable goods of which the natives are poſſeſſed; for, independendependently of their cotton, which in all the interior nations is deſcribed as of common manufacture, and therefore as of general growth, their mines of gold (the improveable poſſeſſion of many of the Inland States) will furniſh, to an unknown, and probably boundleſs extent, an article that commands, in all the markets of the civilized world, a conſtant and unlimited ſale.

Such are the important objects of mercantile purſuit, which the various intelligence obtained by the firſt efforts of the Aſſociation, enables them to point out and recommend to their Country: and while the contemplation of national intereſts, and of the ſtill more extended intereſts of philoſophy, directs their efforts and animates their hopes, they cannot be indifferent to the reflection, that in the purſuit of theſe advantages, and by means as peaceable as the purpofes are juft, the conveniencies of civil life, the benefits of the mechanic and manufacturing arts, the attainments of fcience, the energies of the cultivated mind, and the elevation of the human character, may in fome degree be imparted to nations hitherto configned to hopelefs barbarifm and uniform contempt.

FOR the following Memoir, and for the Map which it describes, the Society are indebted to the eminent talents and ardent zeal in the Promotion of Geographical Improvement that distinguish the character of MAJOR RENNELL.

CHAPTER XII.

Conſtruction of the Map of Africa.

THAT the Geography of Africa has made a ſlower progreſs towards improvement than that of every other part of the world, during the laſt, and the preſent century, is to be attributed more to natural cauſes, than to an abſolute want of attention on the part of Geographers. Formed by the Creator, with a contour and ſurface totally unlike the other Continents, its interior parts elude all nautic reſearch; whilſt the wars and commerce in which Europeans have taken

ken

en part, have been confined to very circumfcribed parts of its borders. Thefe moft productive means of geographical information failing, the next refource is to collect materials from the beft informed amongft the travelled *natives*: I fay *natives*, becaufe the generality of European Travellers reckon upon fome degree of folace, as well as the gratification of curiofity, during their peregrinations: not to mention, that it is more the practice to fee what has been already feen, than to ftrike into a new path, and dare to contemplate an unfafhonable fubject. To the lovers of adventure and novelty, Africa difplays a moft ample field: but the qualification of local manners, and, in fome degree, of habits, muft in this cafe, be fuperadded to that of language: and this, unqueftionably, renders the undertaking more arduous than that of an ordinary Tour. But the Adven-

Adventurer in queſt of fame, will readily appreciate the degree of glory attendant on each purſuit.

The 18th century has ſmiled propitiouſly on the Science of GEOGRAPHY throughout the globe; and an Engliſhman may be allowed to pride himſelf, that his countrymen have had their full ſhare of the glory attending this, and other kinds of reſearches calculated to increaſe the general ſtock of knowledge. It is to this ſpirit that we are to attribute the acquiſition of the materials which form the ſubject of the preſent Work. It is no leſs to this ſpirit that we are indebted for the progreſſive improvements in the North-American and Aſiatic Geography: our ſyſtems embracing objects far ſuperior to thoſe, for which Geographical Surveys are ordinarily undertaken: not the topography of townſhips, diſtricts,

diſtricts, counties; but the Geography of Empires, Regions, and Continents!

As both EUROPE, and its adjacent Continent, ASIA, are ſpread over with inland ſeas, lakes, or rivers of the moſt extended navigations, ſo as collectively to aid the tranſport of bulky articles of merchandize from one extreme of them to the other; and to form (like ſtepping-ſtones over a brook) a more commodious communication: ſo likewiſe the northern part of the new Continent appears to have an almoſt continuous Inland Navigation; which muſt prove of infinite advantage to its inhabitants, when fully peopled; and contribute to their ſpeedier civilization, in the mean time. But Africa ſtands alone in a geographical view! Penetrated by no inland ſeas, like the Mediterranean, Baltic, or Hudſon's Bay; nor overſpread with extenſive lakes,

lakes, like thoſe of North America: nor having in common with the other Continents, rivers running from the center to the extremities; but, on the contrary, its regions ſeparated from each other by the leaſt practicable of all boundaries, arid Deſarts of ſuch formidable extent, as to threaten thoſe who traverſe them, with the moſt horrible of all deaths, that ariſing from thirſt! Placed in ſuch circumſtances, can we be ſurpriſed either at our ignorance of its Interior Part, or of the tardy progreſs of civilization in it? Poſſibly, the difficulty of conveying merchandize to the coaſts, under the above circumſtances, may have given riſe to the *traffic* in *men*, a commodity that can tranſport itſelf! But laying this out of the queſtion, as an abſtract ſpeculation, there can be little doubt but that the progreſs of civilization amongſt the Africans, has been

as

as ſlow as can be conceived, in any ſituation: and it has alſo happened, of courſe, that the deſtined inſtruments of their civilization, have remained in a proportional degree of ignorance concerning the nature of the country.

Nothing can evince the low ſtate of the African Geography, more than M. D'Anville's having had recourſe to the Works of PTOLEMY and EDRISI, to compoſe the Interior Part of his Map of Africa (1749.) It is well known, that thoſe Authors wrote in the ſecond, and in the twelfth centuries of our æra. Moſt of the poſitions in the Inland Part of the great body of Africa are derived from EDRISI; and it is wonderful how nearly ſome of the poſitions agree with thoſe furniſhed by the preſent materials. Such was the tranſcendant judgment of D'Anville, in combining the ſcanty notices

tices that are furniſhed by the Nubian Geographer!

But the Public are not to expect, even under an improved ſyſtem of African Geography, that the Interior Part of that Continent will exhibit an aſpect ſimilar to the others; rich in variety; each region aſſuming a diſtinct character. On the contrary, it will be meagre and vacant in the extreme. The dreary expanſes of deſart which often ſurround the habitable ſpots, forbid the appearance of the uſual proportion of towns; ard the paucity of rivers, added to their being either abſorbed or evaporated, inſtead of being conducted in flowing lines to the ocean, will give a ſingular caſt to its hydrography; the direction of their courſes, being moreover, equivocal, through the want of that information, which a communication with the ſea uſually affords at at a glance. Little as the Antients knew of the Interior Part of Africa, they appear to have underſtood the character of its ſurface; one of them comparing it to a leopard's ſkin. Swift alſo, who miſſes no opportunity of being witty at the expence of mathematicians, diverts himſelf and his readers both with the *nakedneſs* of the land, and the abſurdity of the map-makers.

> "Geographers, in *Afric* maps,
> With ſavage pictures fill their gaps,
> And o'er unhabitable downs
> Place elephants, for want of towns."

The Society for Promoting the Diſcovery of the Interior Parts of Africa has been fortunate in collecting much geographical information, in ſo early a ſtage of the African reſearches; and there is little doubt but that in a few years all the great

great features of this Continent (within the reach of their enquiries) may be known and defcribed. But to accomplifh this, it will be neceffary that intelligent Europeans fhould trace fome of the principal routes; as well to apportion the diftances, as to eftablifh fome kind of criterion for the parole information derived from the natives. As yet, in the wide extent of near thirty degrees on a meridian, between BENIN and TRIPOLI, not one celeftial obfervation has been taken, to determine the latitude.

But as far as the materials compofed by a fcale founded on eftimated diftances (that is days journeys of caravans) may go towards the eftablifhment of geographical pofitions, the itineraries made ufe of for conftructing the new matter in the accompanying map, are lefs difcordant than might be expected in fo wide

an

an expanſe, and on the foundation of rules ſo vague as thoſe which neceſſity has compelled me to apply. On this, however, the Reader muſt exerciſe his own judgment, after the following expoſition of the *data*, and the rules by which I have determined the ſcale.

It will eaſily be conceived by the Reader, that all roads, except ſuch as are made through a country, in which the public works are in the higheſt ſtate of improvement, and where alſo the face of it is perfectly level (the curvature of the earth excepted) muſt have ſome degree of inflexion, both horizontally and vertically; and either of theſe will occaſion a Traveller to trace a line of greater length by the road, than can be meaſured on a ſtraight line, drawn from the point of commencement, to that of the termination of his journey. The quantity of the

the difference muſt vary with the nature of the country; but in ordinary caſes, ſtill more with the extent of the line of diſtance: for a different ratio between the road diſtance and horizontal diſtance muſt obtain, as the line of diſtance is increaſed. Let it be admitted, that in ſtages of ten or twelve miles, the winding of the road occaſions a loſs of only one-tenth part, which may be termed the *ſimple* winding: yet as the different ſtages in an extent of 100 or 150 miles, do not lie in a ſtraight line, drawn through the whole extent, but often very far to the right and left of it, a *compound* winding ariſes: and I have found by long experience, that one mile in eight muſt be deducted, to reduce the road meaſure, on ſuch a length, to horizontal meaſure. When a line of diſtance is extended to 500 miles and upwards, the rule becomes much more vague than when applied to moderate diſtances; becauſe it often happens (and more particularly in unimproved countries) that obſtacles ariſe, which give an entire new direction to the courſe of the road; although the two parts of it, conſidered ſeparately, may have only an ordinary degree of crookedneſs. However, as ſome of the lines of diſtance applied to the preſent ſubject, are from thirty to forty days journey, it becomes neceſſary that ſome general rules ſhould be adopted. It happens that examples are furniſhed, in two caſes, on very long journies, where the real diſtances between the terminating points of the routes are nearly known: ſuch is that of fifty-three days journey, between the Capital of *Fezzan* and *Cairo*; and the mean horizontal diſtance for each day, is fourteen and a half geographic miles, or thoſe of ſixty to a degree. I confeſs

I ſhould

I ſhould have expected much leſs. The other example is between *Arguin* and *Gallam*: there forty days produce thirteen miles for each day; and this is conformable to my expectations. In the examples of ſmall diſtances, ſuch as ſix days journey, ſixteen miles per day is the reſult, and is conſiſtent. For a caravan journey, taken at twenty-two Britiſh miles of *road* diſtance, will produce, when the allowance for winding is deducted, and the remainder reduced to geographic miles, about ſixteen and a half ſuch miles for a ſingle day.

The following are the proportions which I have eſtabliſhed, for the application of a ſcale, to the different degrees of diſtance.

For one day, ſixteen miles and a half; for ſeventeen to twenty-five days, fifteen miles; for forty to fifty days, thirteen miles. Theſe numbers are particularly ſelected, becauſe they occurred in the courſe of the Work. The Reader will be pleaſed to obſerve, that the miles ſpoken of in the conſtruction, are always thoſe of ſixty to a degree of a great circle. However tedious this inveſtigation may appear to the generality of readers, it is abſolutely neceſſary; as it is the hinge upon which the whole turns: and a neglect of attention to this particular ſubject, would warrant the Reader's taking the whole for granted, without further examination.

Mr. BEAUFOY having given, from the materials in his poſſeſſion, ſo full an account of each road and country, nothing remains for me to do, but ſimply to deſcribe, from the ſame materials, the mode of fixing the principal poſitions, in the Map. As the object of it is to exhibit the new matter only, care has been taken to exclude all that has already appeared, except what was abſolutely neceſſary towards explaining the other: and as the borrowed particulars are diſtinguiſhed from the reſt, the Reader cannot be miſtaken. The outline of the great body of Africa, together with the courſes of the Nile, Gambia, Senegal, and Wad-drah, are copied from Mr. D'Anville.

Fezzan (or rather its capital, Mourzouk) is given in the Itineraries at the diſtance of ſeventeen days and a half from Meſurata. Theſe, taken at fifteen miles per day, produce two hundred and ſixty-two miles. The bearing is ſaid to be South from Meſurata; and this latter is placed according to D'Anville. Mourzouk, then, falls in latitude 27° 20′.

Agadez, the next principal ſtation, is, at a medium forty-one days from Mourzouk, on a South Weſt courſe, or thereabouts: and theſe, at thirteen miles per day, produce four hundred and fifty-five miles; and place Agadez in latitude 20° 20′; and nearly in the meridian of Tripoli. Agadez is the Agadoſt of Edriſi.

From Agadez to Caſhna is ſeventeen days; which, at fifteen miles per day, give a diſtance of two hundred and fifty-five miles. The bearing is ſaid to be South South Weſt. Caſhna, then, will ſtand in or about the latitude of 16° 20′ North, and about a degree and a half Weſt from Tripoli.

D'Anville's *Caſſeenah* (undoubtedly meant for Caſhna) is placed about thirty-ſeven miles to the North Weſt of the poſition

poſition aſſumed in the accompanying Map; whence I conſider mine as a near approximation, eſpecially as the diſtance from Meſurata is upwards of nine hundred and ſeventy miles.

Caſhna may be regarded as the central kingdom of the great body of Africa; and as a part of the region named SOUDAN, of which at preſent but few particulars are known.

Ghanah, or Ghinnah, is placed, in reſpect of Caſhna, according to M. D'Anville's Map. It does not appear whether he had any authority for placing it ninety miles to the North Eaſt of Caſhna: but its poſition, in reſpect of the City of Nubia, (antiently Meroé, on the Nile) is on the authority of Edriſi. This Author alſo allows twelve days between Agadez and Ghanah: and by my conſtruction, they are two hundred and eight miles aſunder. See *Geog. Nubienſis*, p. 39. Ghanah was in the twelfth century a city of the firſt conſequence. Wangara and Kanem, were alſo known to Edriſi.

The river known to Europeans by the name of NIGER, runs on the South of the kingdom of Caſhna, in its courſe towards Tombuctou; and if the report which Ben Alli heard in that town, may be credited, it is afterwards loſt in the ſands on the South of the country of Tombuctou. In the Map, only the known part of its courſe is marked by a line; and the ſuppoſitious part by dots. It may be proper to obſerve, that the Africans have two names for this river; that is, NEEL IL ABEED, or RIVER of the NEGROS; and NEEL IL KIBEER, or THE GREAT RIVER. They alſo term the

NILE,

NILE, (that is the Egyptian River) NEEL SHEM: ſo that the term NEEL, from whence our NILE, is nothing more than the appellative of River; like GANGES, or SINDE.

From Caſhna the road leads Weſtward to the Kingdom of GONJAH, ninety-ſeven days journey from the former. *Gonjah*, is, from circumſtances, the *Conché* of M. D'Anville, and the *Gongé* of M. Deliſle; and the ſimilitude of names, however great, is the leaſt proof of it: for the Itinerary of the SHEREEF IMHAMMED ſays, that eighteen or twenty days from Gonjah, towards the North Weſt (or between the Weſt and North) lies the Country of YARBA: and eight days farther Weſt, that of AFFOW. Now the countries of YARRA and YAFFON, will be found in Deliſle's Map of Senegal

(1726), nearly in the poſition that *Yarba* and *Affow* take in reſpect of Gonjah; ſuppoſing D'Anville's *Conché* to be meant for it. It is extremely difficult to aſſign a ratio for the decreaſe of the horizontal diſtance, on ſo extended a line as ninety-ſeven days journey; and therefore it would be loſing time to attempt it. Gonjah, by circumſtances, is about eight hundred and ſeventy miles from Caſhna; which allows only nine miles for each day. I therefore conclude that the road is very circuitous.

Gonjah is reported by the Shereef to be forty-ſix days journey from the Coaſt of Guinea, to which the Chriſtians trade. It is probable that the Gold Coaſt is the part meant, and that may be taken at five hundred and thirty miles from D'Anville's Conché. The ratio at thirteen per day,

day, would give near ſix hundred. Here again, it would be loſing of time, to reaſon on ſuch a point of uncertainty, ſince neither of the extreme points of the line of diſtance are correctly known. The Reader muſt therefore determine for himſelf. Of this ſpace of forty-ſix days travelling, from Gonjah towards the Coaſt, the Shereef had travelled only the firſt ten days, to the City of Kalanſhee, a dependecy of the Kingdom of *Tounouwah*; the capital of which, according to his report, ASSENTAÍ (the Aſſianté of D'Anville) ſituated midway between Kalanſhee and the ſea coaſt: that is, eighteen days journey from each. The Shereef alſo reports, that there is no communication between this coaſt (which we may ſuppoſe to be the Gold Coaſt) and the country of Gonjah: for that the King of Aſſentaí, who poſſeſſes the ſpace

tween,

betwen, prohibits his Inland Neighbours from paſſing through his country.

But Mr. NORRIS, a gentleman who reſided many years in Wydah, &c. reports differently: for he ſays, that there are other States, (that is, the Fantees, and their confederates) lying between Aſſentaï and the ſea; and that the Aſſentais have often attempted, but without ſucceſs, to open a communication with the Coaſt.

To return to the route from Caſhna to Gonjah. There are between them ſome extenſive kingdoms or ſtates, moſt of which appear to preſerve their antient religion. I have generally marked the progreſs of the Mahomedan Religion, by a creſcent; and the Caffre States by an arbitrary mark of a different kind.

It will appear by a ſlight inſpection of the Map, that the Mahomedan Religion, as far as reſpects the Interior Part of the Country, has ſpread ſouthward, to about the parallel of twelve degrees of North latitude. Probably though, in ſome of thoſe countries where the Court religion is Mahomedan, the bulk of the people may profeſs the antient religion.

TOMBUCTOU, is placed on the following authorities: Firſt, Mr. MATRA, the Britiſh Conſul in the dominions of Morocco, ſays, on the authority of the natives, that Tombuctou is fifty days *caravan travelling* from TATTAH, a place ſituated on the common frontiers of Morocco, Drah, and Zenhaga; and in the route from Morocco, and Suz, to Tombuctou. Tattah is aſcertained in poſition, by a route of Ben Alli's. He found it to be nine days and half from Moroc-

co,

co, and one day ſhort of a ſtation on the Wad-drah (or Drah River) which ſtation was four days, or ſixty-ſix miles lower down than *Tinjuleen*, a place in D'Anville's and Deliſle's Maps of Africa. It was alſo twelve days journey from the City of Nun, or Non, which city by Ben Alli's account, is two days from the ſea coaſt; and well known to be oppoſite to the Cape of the ſame name. Theſe authorities enable us to place Tattah one hundred and ſeventy miles South South Eaſt from Morocco. Then, fifty days from Tattah to Tombuctou, at thirteen each day, produces ſix hundred and fifty miles. By Ben Alli's report, Tombuctou is forty-eight days from the capital of Sultan Fullan, lying within the diſtrict of Gallam, on the River Senegal. The poſition of this place is not known to me; but by circumſtances it muſt be near the river: and in uſing materials of ſo

coarſe

coarſe a kind, trifles muſt not be regarded. Forty-eight days at thirteen each day, produce ſix hundred and twenty miles; and this line of diſtance meets that from Tattah, in latitude 19°. 40′. and nearly midway between Gallam and Caſhna. In this poſition, it falls only twenty-eight miles to the North Weſt of D'Anville's Tombuctou.

It appears that moſt of the road from Tattah to Tombuctou, lies acroſs the vaſt Deſart, commonly known by the name of ZAHARA, or properly, THE DESART. Geography is at preſent, very bare of particulars, in this quarter. Ben Alli went from Tombuctou, direct to Fezzan, ſkirting the South Eaſt border of this great Deſart. He reckoned only ſixty-four days between Tombuctou and Fezzan, which at twelve miles and half per day, produce only eight hundred miles.

miles. The interval on the Map is nine hundred and ſeventy. Reaſon, however, points out, that the diſtance from the neareſt place, Gallam, ought to be preferred. And as it is underſtood, that Agadez and Tombuctou are about fifty-five days aſunder, it appears yet more probable that the interval between Fezzan and Tombuctou, ought not to be reduced. It muſt be recollected, that Ben Alli's Communications were given from memory, after an interval of twenty years.

The point of the next importance, is Bornou, the capital of an extenſive kingdom ſituated on the South Eaſt of Fezzan, and between the two NEELS or NILES; that of Egypt, and that of the Negros.

Bornou, is given by the Shereef, at about

about fifty days from Mourzouk (or Fezzan) which may be taken at six hundred and fifty miles. He also reports that it is twenty-five days journey from the course of the Nile, where it passes the country of Sennar; or in distance about three hundred and sixty miles. This would place Bornou in a direction of South East, somewhat southwardly, from Mourzouk; and about the parallel of 19° 40′. It is not probable that Bornou has a more westerly position. Edrisi's account would place it more easterly; for he says, *Matthan*, a city of *Kanem*, lies thirty-one days from *Nuabia* (on the Nile) through *Tegua*. *Geog. Nub.* p. 15. Edrisi's day's journey is equal to eighteen Arabic miles, or nearly nineteen geographic ones: consequently the thirty-one days give five hundred and eighty-seven miles. Matthan is not reported in the Itinerary: but Kanem is,

 both

both as Province and a capital City: and the poſition of the latter according to my conſtruction, is ſeven hundred miles from Nuabia. Whether the error lies on the ſide of Edriſi, or the Shereef; or ariſes from the faulty poſition of Nuabia in D'Anville, cannot eaſily be diſcovered.

Ben Alli travelled the road between Bornou and Alexandria; but was too much indiſpoſed to make any obſervations, otherwiſe than that the Kingdom of Bornou extends fifteen days journey, or about two hundred and thirty miles, in that direction. This was particularly unfortunate; for whichſoever route he went, he muſt probably have croſſed ſome one or more of the OASES; and of courſe ſome important matter would have been added to the Map.

The

The only route of importance that remains to be difcuffed, is that from Fezzan or Mourzouk, to Cairo, leading to Mecca; for, at Mourzouk the Mahomedans from the fouthern and weftern parts of Africa, who intend to make a pilgrimage to Mecca, affemble at the proper feafon, as at a common point of departure. The route to Cairo, which requires fifty-three days to atchieve, appears on the Map to be feven hundred and feventy miles; being equal to fourteen and half for each day: and is on the whole, feventy more than I fhould have expected that number of days to produce. To what degree of accuracy the difference of longitude between Mefurata and Cairo, is afcertained, I am ignorant: as alfo whether the bearing of Mourzouk from Mefurata, be right. Fourteen miles and half of horizontal diftance for each day, on fo long a line of diſtance, and on ſo rugged a way as the Itinerary deſcribes, is too great a proportion; and we may ſuſpect an error ſomewhere.

A circumſtance occurs in the Itinerary, which would determine how near this route approaches to the Coaſt of the Mediterranean Sea; if we might depend on the accuracy of the Itineriſt. The dates produced in the Plain of *Gegabib*, are gathered by the people of *Derna*, who are ſaid to inhabit the ſea coaſt, eight days journey off; or about one hundred and thirty miles. *Derna* (antiently *Darnis)* is ſituated within the confines of Tripoli: but the diſtance muſt be faulty; becauſe *Augela* is the neareſt point in this route, to Derna, though ten days from it: and *Gegabib*, is ſeven days from Augela, in a direction that ſtill increaſes the diſtance.

Augela is found in Herodotus, Book IV. under the name of *Ægila*; and in Ptolemy and Pliny it is written *Augila*. In Ptolomy, (Africa, Tab. III.) it is placed about one hundred and ninety-ſeven miles from the ſea coaſt, and about a degree of longitude to the eaſtward of *Darnis*. Its longitudinal poſition from Mourzouk and Cairo, agrees very well: and conſidering the extent of Ptolemy's local knowledge in this quarter, we may ſuppoſe him well acquainted with its diſtance from the coaſt. Allowance muſt be made for an exceſs of diſtance given by Ptolemy's ſcale, in this Map; and it being in the proportion of twenty-ſeven to twenty-three, the one hundred and ninety-ſeven miles ſhould be reduced to one hundred and ſixty-eight: and according to this, Augela ought to ſtand in latitude 29° 20′; and nearly midway between Mourzouk and Cairo.

Gegabib, as has been ſaid before, is ſeven days journey from Augela, towards Cairo; and as I have a particular pleaſure in producing any authority that ſerves to prove the veracity of ſuch an Author as HERODOTUS, I ſhall juſt mention, that (in B. IV.) he ſays, that the *Naſamones* in the Summer ſeaſon, leave their cattle on the *coaſt*, and go to the plains of Ægila, to gather the fruit of the Palm trees, which abound in that place. And he marks the poſition of this coaſt, by placing it on the Weſt of *Teuchera*, or *Tauchira*, (the modern *Teukera*) a ſea-port that lay within the diſtrict of *Cyrene*, now better known by the names of Curin and Barca.

But

But Ptolemy's Naſamones occupy, in reſpect of *Augila*, the very ſpot where the dates are now gathered in the plain of *Gegabib:* and therefore their territories muſt have extended at leaſt from that plain, to the eaſtern coaſt of the Great Syrtis. And although the preſent gatherers of the Augela dates may come from different parts of the coaſt, from thoſe, in the time of Herodotus; yet it is remarkable, that the ſame cuſtom ſhould have exiſted for twenty-two or twenty-three centuries.

Between *Augela* and *See-wah*, the next town towards Cairo (and probably the *Siropum* of Ptolemy) the road paſſes over a chain of very high mountains, named in theſe times *Gerdobah:* and this is unqueſtionably the ſame ridge that terminates on the Mediterranean, a few days journey farther on; and which, by the ſuddenneſs of its deſcent towards the ſea, was antiently named *Catabathmus*. This chain or ridge divided *Cyrene* from *Marmarica*.

The ſite of the antient Temple of *Jupiter Ammon*, was a few days journey (perhaps four or five) inland from the plain of Gegabib, ſo often mentioned. I think I may venture to ſay this, on the authority of Herodotus, Strabo, Pliny, and Arrian; from each of whom, ſome particulars may be collected reſpecting its ſituation. Firſt, Herodotus ſays, (Book IV.) that the Temple is ſituated ten days from *Ægila*; (frequented by the Naſamones on account of the dates) and on the road from Thebes to *Ægila*. Next, Arrian ſays, on the authority of Ariſtobulus, that Alexander went to it, from the ſite of his new city of Alexandria, along the ſea coaſts of Egypt and Marmarica,

marica, to *Paraetonium:* which latter was ſituated, according to the ſame authority, ſixteen hundred ſtadia from Alexandria. Pliny gives it at two hundred Roman miles (Book V.) Theſe accounts are perfectly concluſive; and the poſition of *Paraetonium*, is alſo very clear in Ptolemy; and is moreover known to the Moderns under the name of *Al Bareton:* ſo that no difficulty can ariſe here. Arrian ſays farther, that Alexander ſtruck inland from *Paraetonium*, and entered the Deſart: but he does not ſay how far the Temple lay from the ſea coaſt. This is ſupplied by Strabo, (Book XXVII.) who gives the diſtance at thirteen hundred ſtadia. Allowing theſe to produce one hundred and thirty, or one hundred and forty miles; and taking Herodotus's ten days from Augela, at one hundred and ſeventy (we muſt not conſider them as *caravan* journies, but as *ordinary* ones) the

meeting

meeting of theſe lines of diſtance, place the Temple in latitude twenty-nine degrees, and a ſmall fraction; and in a South Weſterly direction from Paraetonium. Pliny ſays, (Book V.) that the Temple is four hundred [Roman] miles from Cyrene; that is, twice as far as Paraetonium is from Alexandria: and this agrees with the former poſition. Laſtly, Ptolemy places it one hundred and ninety-five geographic miles from *Paraetonium*; and from *Cyrene* three hundred and forty.—But Ptolemy's ſcale, in Africa, Tab. III. gives too much diſtance (as I have ſaid before) and corrected, it ſhould be one hundred and ſixty-ſix from Paraetonium. As theſe authorities do not vary amongſt themſelves more than thirty miles, I conſider them concluſive.

M. D'Anville's poſition of this Temple is about thirty miles farther to the Southward;

ward; that is, from the Mediterranean; but he does not quote his authorities. In his *Geographie Ancienne Abregee*, vol. iii. p. 42. he has the following paſſage: "Selon la Géographie actuelle, ce qu'on "trouve ſous le nom de *Santrieh*, paroit "en tenir la place; & par la Nature du "pays, qui ne laiſſe point diſtinguer "d'autre objet, on n'eſt point embar-"raſſé ſur le choix." Edriſi *(Geo. Nub.* p. 41.*)* places *Santrie* ten days Eaſtward from Augela, and nine days from the Mediterranean; which carries Sant-rie farther from the ſea coaſt, than Strabo allows to the Temple; but accords with Ptolemy. Savary (Vol. ii. Lett. VIII.) quotes Abulfeda, to ſhew that the *Oaſes* were only three days journey Weſt of the Nile; and Ptolemy places the largeſt of them, named EL WAH by the Arabs, under the parallel of 27°. I ſuſpect Abulfeda is wrong; and that Ptolemy is

nearer

nearer the truth, when he allows one hundred and twenty-three miles, (or one hundred and five corrected) for the diſtance of the Great Oaſis from Ptolemais on the Nile, in the direction of Weſt, ſomewhat Southwardly. Then Edriſi allows only nine days between Sant-rie and El Wah; whereas the ſite of the Temple of Jupiter Ammon, by the above authorities, ſhould be, according to my apprehenſion, at leaſt twelve days from EL WAH. But we are young in African Geography: and as I have ſaid before, the *data* furniſhed by Arrian, Strabo, and Pliny, may ſatisfy us.

The deſcription of the *Oaſis* (or Iſland in the midſt of the ſandy Deſart) which contained the Temple, is pretty generally known: but for the ſake of thoſe who may not recollect the particulars, I have extracted the following account.

Arrian

Arrian ſays, that it is not more than forty ſtadia in extent; Diodorus fifty: ſay, ſix or ſeven miles. All accounts agree, that it has one or more fountains of water; and that it was planted with divers kinds of fruit trees: Arrian particularly notices the Palm and Olive. What appeared to be a very great natural curioſity, was, a fountain, which according to Arrian, (whoſe account is the leaſt extravagant) varied in its temperature, in a greater degree than any other that has been heard of: that is, it was very warm, or hot, at midnight; very cold in the heat of the day. I preſume theſe phænomena will not appear very extraordinary to thoſe, who conſider, that a deep-ſeated ſpring will preſerve a mean degree of temperature at all ſeaſons: ſo that, in effect, it was the atmoſphere that underwent the change; and with

it,

it, the bodies of thoſe who made the obſervations.

The Temple was ſurrounded by a triple wall, forming three diſtinct quarters or diviſions; one of which was appropriated to the uſe of the Monarch. In the time of Herodotus, when probably the Temple was in its glory, the dominions of the Ammonites reached within ten days journey of the City of Thebes: the people were a colony of Egyptians and Ethiopians, and ſpoke a mixed language, (Herod. Book III.) Ammon, or Hammon, was the Egyptian name of Jupiter; and the image of the god, ſimilar to that at Thebes; that is, it had the head of a Ram. (Book IV.)

In the time of Strabo, about four hundred and fifty years after Herodotus, the Temple

Temple was almoſt deſerted; as the Oracle was grown out of faſhion.

It is probable that ſome remains, either of the triple wall, or of the Temple, may be found at this day; although the materials may have undergone a different kind of arrangement. The tranſport of the materials acroſs the Deſart could only have been accompliſhed by the ſtrong impulſe of ſurperſtition: and being once collected, nothing but a like cauſe could remove them. See-wah appears to be the neareſt town to this Oaſis; and is probably not more than ſix days journey on the North Eaſt of it: the ſpring, together with the ruins of the Temple, and the triple wall, might aſcertain the ſpot, if the curioſity of the preſent age demanded it.

MARCH, 1790.